Estrada Ramírez

Energy and Productive Efficiency in Polymer Processing

Omar Augusto Estrada Ramírez

Energy and Productive Efficiency in Polymer Processing

A New Practical Approach with Real-Case Applications

HANSER

Print-ISBN: 978-1-56990-414-5
E-Book-ISBN: 978-1-56990-948-5

Bibliographic information of the German National Library:
The German National Library lists this publication in the German National Bibliography; detailed bibliographic
data are available on the Internet at http://dnb.d-nb.de.

© 2025 Carl Hanser Verlag GmbH & Co. KG, Munich
Vilshofener Straße 10 | 81679 Munich | info@hanser.de
www.hanserpublications.com
www.hanser-fachbuch.de
Editor: Dr. Mark Smith
Production Management: Eberl & Koesel Studio, Kempten
Cover concept: Marc Müller-Bremer, www.rebranding.de, Munich
Cover design: Max Kostopoulos
Cover picture: AdobeStock/molpix; Thomas West
Typesetting: le-tex publishing services GmbH, Leipzig

This book is dedicated to Liliana, my wife, who has had the courage to persevere despite everything. It is also dedicated to three very special people whom I consider my "senseis" and to whom I owe much of my professional development: Dr. María del Pilar Noriega, Dr. Farid Chejne, and Dr. Robin Kent.

The Authors

The Editor

Omar Augusto Estrada Ramírez, Ph.D., is CEO and founding partner of SOSPOL S.A.S., CEO and founding partner of QUI-MERA PLASTICS S.A.S., and Partner at PCR-GREEN S.A.S. He is also External Senior Researcher at the ICIPC, Medellín, Colombia, and Senior Researcher in the Colombian National System of Science and Technology.

Formerly, he was Technical and Scientific Director of the IC-IPC. He achieved his chemical engineering degree and his Doctor of Engineering with an emphasis on energy systems from the National University of Colombia – Faculty of Mines.

He achieved a degree as Specialist in Plastic and Rubber Transformation Processes and a Master's in Polymer Processing Engineering from the joint programs between EAFIT University and the ICIPC. He was Coordinator of the Specialization in Plastic and Rubber Transformation Processes for nine cohorts.

He is the creator of the Energy Gaps Method for the diagnosis and identification of intervention priorities to improve energy and production efficiency in polymer processing plants, having successfully intervened in more than 40 production lines in Colombia. As a specialist in advanced analytics of energy consumption and production data, he has developed tools that allow companies to obtain greater benefits from existing information and monitoring systems.

As a specialist in plastic materials, polymer processing by extrusion, blow molding, and thermoforming, energy and production efficiency in polymer processing, and sustainability of the plastics industry, he has advised more than 100 companies throughout his professional career, successfully participated in more than 25 research, development, and innovation projects, and has given more than 250 conference papers/presentations and open and closed courses on different topics of his specialty.

The Contributors

Chapter	Title	Authors
1	Introduction	Omar Estrada[a], Farid Chejne[b]
2	Practical Concepts of Energy in Polymer Processing: A View from Macroscopic Energy Balances	Omar Estrada[a], Farid Chejne[b], Juan Carlos Maya[c]
3	Monitoring and Targeting (M&T)	Julián Patiño[d], Omar Estrada[a], Farid Chejne[b]
4	The Energy Gap Method (EGM)	Omar Estrada[a], Iván López[e]
5	Case Studies of Energy Optimization in Polymer Processing Using the EGM	Nicolás Muñoz[f], Omar Estrada[a], Iván López[e]
6	Industry 4.0 Enabling Technologies	Julián Patiño[d], Carlos Correa[g], Omar Estrada[a]

a PhD Omar Augusto Estrada Ramírez. ICIPC - Instituto de Capacitación e Investigación del Plástico y del Caucho y SOSPOL - Soluciones en Sostenibilidad y Polímeros S.A.S. E-mail: *oestrada@sospol.co*

b PhD Farid Chejne Janna. Universidad Nacional de Colombia. E-mail: *fchejne@unal.edu.co*

c PhD Juan Carlos Maya López. Universidad Nacional de Colombia / Sume EnergyC S.A.S. E-mail: *jcmaya@unal.edu.co*

d PhD Julián Alberto Patiño Murillo. ICIPC - Instituto de Capacitación e Investigación del Plástico y del Caucho. E-mail: *jpatino@icipc.org*

e PhD Iván Darío López Gómez. SPE - Society of Plastics Engineers and SOSPOL - Soluciones en Sostenibilidad y Polímeros S.A.S. E-mail: *ilopez@sospol.co*

f MSc. Nicolás Muñoz Realpe. ICIPC - Instituto de Capacitación e Investigación del Plástico y del Caucho. E-mail: *nrealpe@icipc.org*

g MSc. Carlos Mario Correa. Sume EnergyC S.A.S. E-mail: *cmcorrea4@gmail.com*

Foreword

It gives me great pleasure to write the foreword to this book by my friend and colleague, Omar Augusto Estrada Ramírez. I am fortunate to have had the pleasure of working, corresponding and debating with Omar on energy management and production efficiency in plastics processing since 2018. During this time, I have followed his work and his development and use of the energy gap method from what was initially (to me) an interesting concept into what is now a fully-fledged tool for diagnosing energy and production issues in plastics processing. Not only does the energy gap method allow diagnosis of a variety of energy and production issues, but it also gives solid and measurable guidance on improving the performance of plastics processors.

The first chapter reviews the driving forces for energy and process efficiency. These should really not need to be reiterated to any company in the current climate but some plastics processors still have not taken even the most basic actions to reduce energy use and improve productivity. These companies will probably not be purchasing this book but they will also probably not be around in the near future unless they take action. The second chapter introduces the reader to the fundamental concepts of energy use in plastics processing and the third chapter examines the traditional tools and techniques of energy management which have developed over the past 30 years and which I have used in over 550 site surveys across the world. These tools are remarkably effective and can be used to internally and externally benchmark a plastics processor against their previous performance and against similar processors. These are very different benchmarks: internal benchmarking (using Performance Characteristic Lines and CUSUM diagrams) compares a site to what it has previously achieved, whereas external benchmarking (using Performance Characteristic Curves) compares a specific site with other sites using similar processes and simultaneously corrects for production rate variations. Combining internal and external benchmarking allows any site to not only ensure that it is continuously improving but also to ensure that it is achieving world-class standards. However, valuable as this process is, the currently available tools do not always provide a site with the specific knowledge of what it is that is needed to improve or change. This means that it is possible for a site to have evidence that things are not as they should be but to still lack guidance on the required improvement actions. This often requires the use of external consultants who have direct experience of the most profitable actions and the location. The fourth chapter lays the theoretical foundations of the energy gap method by deconstructing the Specific Energy Consumption of a process into five sequential components (Production, Quality, Process, Technology and R&D) with associated energy gaps that make up the total energy used in the process. This segmentation of the total energy use allows a site to use specific tools and techniques to reduce each of the energy gaps and to reduce the total energy used. These tools and techniques are well illustrated with practical examples and suggestions.

Had the book ended there then it would have been a worthwhile contribution to polymer processing literature but Chapter 5 is where the book really starts to get interesting. It takes the previously developed theory and applies it in 10 real-world examples covering injection molding, extrusion (profile, sheet and film) and extrusion blow molding to improve the energy use and process efficiency. Every example illustrates the method with an initial process diagnosis, a process intervention (sometimes two) and a full review of the results of the intervention. This chapter covers a wide range of the factors that affect energy use and productivity and shows how energy gap analysis can be used to look at quality, cycle times, set–up times and all the vital aspects of running plastics processes. This is a major expansion of the theoretical energy gap method into practical plastics processing and every plastics processor can learn from these examples.

Although the development of Industry 4.0 and AI has been slower than predicted in the bulk of the plastics processing industry, the pace is increasing and new machines and technologies are being deployed across the industry to improve energy and production efficiency. These developments form the basis for Chapter 6 where the authors provide a comprehensive review the systems architecture of Industry 4.0 and highlight the challenges and best practices in implementing this. As ever, there is a real–world example of the techniques being used to improve energy and production efficiency.

There is an enviable clarity in all the sections and the ability to achieve clarity is only achieved from a deep understanding of the subject. In summary, I wish that I had read this book early in my career in plastics processing and at this late stage I wish that I had written this book myself.

Dr. Robin Kent

Tangram Technology Ltd.

Hitchin, UK

June 2025

Contents

1

Introduction

Omar Estrada, Farid Chejne

Energy efficiency has been a critical issue since the Industrial Revolution, when the mass production of goods and services significantly increased energy consumption. In the 1790s, Boulton and Watt's steam engines were more fuel-efficient and monetized this increased efficiency by charging a portion of the savings, similar to how performance contracts work today [1]. After World War II, energy efficiency became a significant concern in many countries as reconstruction and economic growth increased energy demand. In the 1970s, the oil crisis led many nations to implement energy efficiency policies to reduce their dependence on imported oil [2]. In the 1980s and 1990s, energy efficiency in industry and the construction sector gained importance, fostering the development of technologies and practices to improve the energy efficiency of buildings and industrial processes [3]. In the 2000s, energy efficiency became institutionalized within global energy policy, with many countries establishing energy efficiency targets and incentive programs to encourage the adoption of more efficient technologies and practices. Today, energy efficiency is a key tool for reducing greenhouse gas emissions and mitigating climate change [4].

1.1 The Importance of Energy and Productive Efficiency

There are at least three reasons why it is essential to focus on energy efficiency in industry: enhancing the competitiveness of companies, reducing greenhouse gas emissions, and the application of practices of corporate social responsibility.

1.1.1 Company Competitiveness

From another perspective, the rising cost of energy significantly impacts product costs. Polymers, characterized by their high calorific value and low thermal conductivity, require substantial energy for melting and cooling. Figure 1.1 illustrates the kilowatt-hours (kWh) needed to generate USD 0.25 of added value in Colombia, showing that the plastic industry requires up to twice as much energy as the manufacturing industry on average to achieve the same added value.

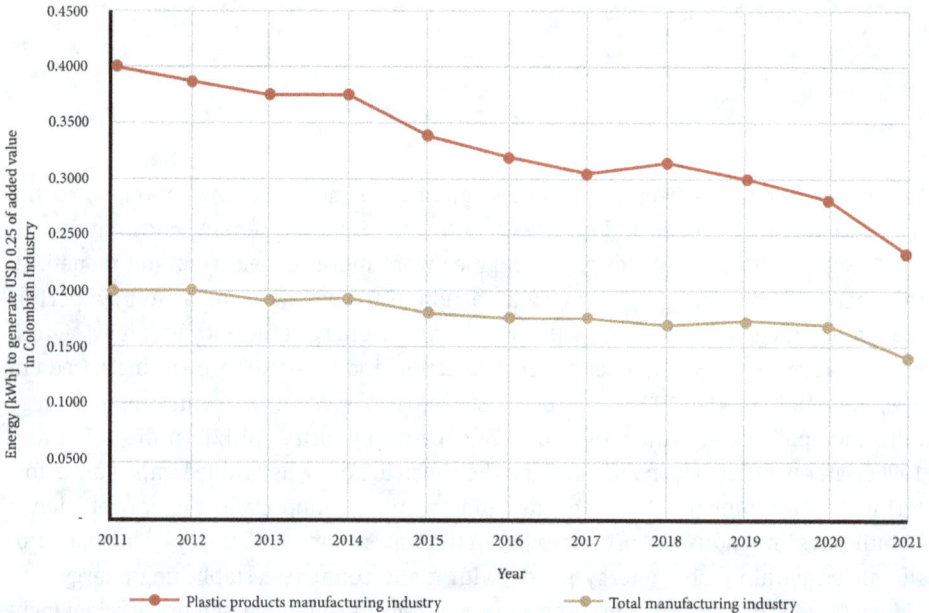

Figure 1.1 Average kilowatt-hours (kWh) required to generate USD 0.25 of added value in Colombia, comparing the plastic industry (red) with the total manufacturing industry (orange), based on data from the Annual Manufacturing Survey by DANE, in Colombia [5]. This highlights the relatively higher energy intensity of the plastic sector

Depending on the country, level of automation, and type of polymer processing, energy can constitute 5–10% of total costs [6]. A 10% reduction in energy consumption can lead to a 0.5–1% increase in business profitability. The plastics industry, especially the packaging industry, represents the largest consumption worldwide and has low net profit margins of 5–10%. An adequate energy efficiency program can generate savings of 10–30% in production costs, which translates into a net profit margin increase of 1–2% [7]. Improving industrial processes for energy efficiency reduces costs and enhances productivity. Energy-efficient processes often yield additional benefits [8], which will be discussed further in Chapter 3. Moreover, studies indicate that employees in companies that focus systematically on energy and productivity efficiency tend to be more empowered, committed, productive, and satisfied, and develop valuable skills beneficial to the company [9].

1.1.2 Reducing Greenhouse Gas Emissions

In 2007, the Intergovernmental Panel on Climate Change (IPCC) presented a report with four scenarios that explore alternative development pathways, covering a wide range of demographic, economic, and technological driving forces and resulting GHG emissions [10]. The report highlighted the relationship between global energy consumption, greenhouse gas emissions, and projected global warming by 2100. According to the study, the Earth's global temperature could increase by almost 6 °C above pre-industrial levels due to the growth of greenhouse gasses, rising from about 35 Gt CO_2 in 2014 to 40 Gt CO_2 in 2060 if we continue to rely on fossil fuel-based energy generation [10]. This scenario could lead to global catastrophe.

In 2012, the International Energy Agency (IEA) first presented the "2DS scenario" as a viable alternative to limit global warming to 2 °C by 2100 [11] and to minimize environmental impacts, thereby preserving most of the planet's ecosystems. To achieve this goal, it is necessary to reach an annual reduction value of 40 Gt CO_2 by 2060, as depicted in Figure 1.2.

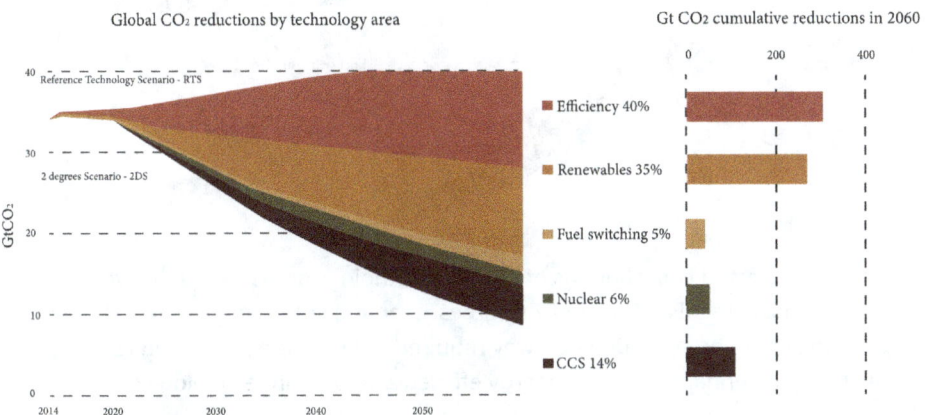

Figure 1.2 Projected contribution of different technology areas to the cumulative reduction of global CO_2 emissions under the IEA's "Updated 2 °C" scenario [12]. The chart reflects expected emission savings up to the year 2060

Figure 1.2 shows that achieving this goal requires energy efficiency to contribute 40% of greenhouse gas emission savings through reduced energy consumption, amounting to a cumulative reduction of 300 Gt CO_2 by 2060. In 1990, physicist Amory Lovins used the term "negawatt" to describe energy saved through efficiency measures [13]. Energy efficiency is also known as the "fifth fuel" because it substantially impacts fuel conservation [14].

However, a special IPCC report first published in 2018 and reprinted in 2022 [15] concluded that limiting global temperature increase to below 1.5 °C is crucial, as higher warming would result in irreversible impacts on critical ecosystems such as coral reefs, the Arctic region, coastal areas vulnerable to flooding, and small-scale fishing communities in low latitudes. At the current rate of temperature increase, this threshold could be reached by 2040. The report also emphasizes that achieving this goal will require unprecedented societal, economic, and technological changes.

From the perspective of reducing greenhouse gas emissions, achieving net emissions of 25 Gt CO_2 by 2030 and net zero emissions by 2060 is essential. While challenging, it is feasible, with energy efficiency expected to contribute 34% more savings than the 2DS scenario. This would increase the expected cumulative energy savings to approximately 400 Gt CO_2 by 2060 in the 1.5DS scenario, up from 300 Gt CO_2 in the 2DS scenario, as illustrated in Figure 1.3.

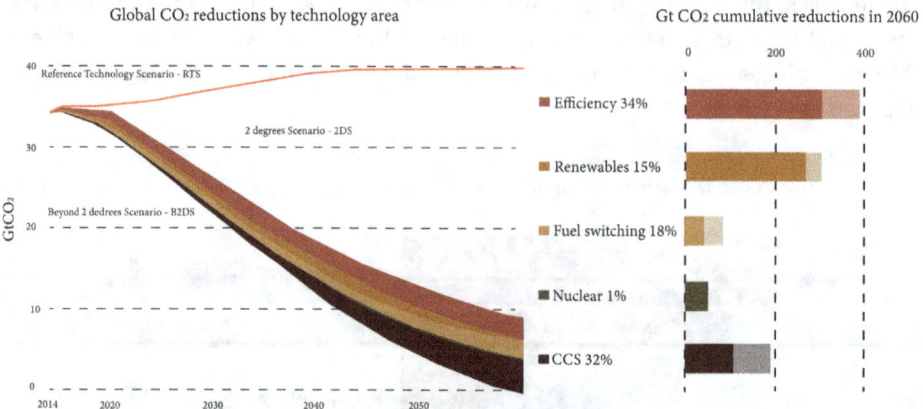

Figure 1.3 Projected contributions of various technology areas to cumulative CO_2 reductions under the IEA's "Beyond 2 Degrees" scenario (1.5 °C target) [11]. The figure outlines the emission reduction pathway required to limit global warming to 1.5 °C, highlighting the enhanced role of energy efficiency and clean technologies

Of the 400 Gt CO_2 that need to be saved globally through energy efficiency, industry is expected to contribute at least 110 Gt CO_2, approximately 27.5% of the expected greenhouse gas emission savings [16]. According to the IEA's 2023 report, global energy consumption currently stands at 442 EJ, with industry accounting for 167 EJ (37.7%) [17]. This report outlines three scenarios that incorporate energy, climate policies, and industrial strategies impacting the role of various technologies in achieving accessible clean energy solutions. These scenarios include the Net Zero Emissions (NZE) scenario, which aims to limit global warming to below 1.5 °C by 2050; the Announced Pledges Scenario (APS), which assumes full implementation of governments' announced commitments on time; and the Stated Policies Scenario (STEP), which reflects the current policy landscape's predominant direction for the energy system.

Figure 1.4 illustrates the projected greenhouse gas emissions trajectories across these scenarios. In 2022, global emissions totaled 37 Gt CO_2. By 2050, under the STEP scenario, emissions would reach 30 Gt CO_2, with global warming projected to reach 2.4 °C by 2100. In contrast, the APS scenario forecasts emissions of 12 Gt CO_2 and global warming of 1.7 °C. Achieving these goals remains challenging, even if countries fully implement their commitments, which remains uncertain.

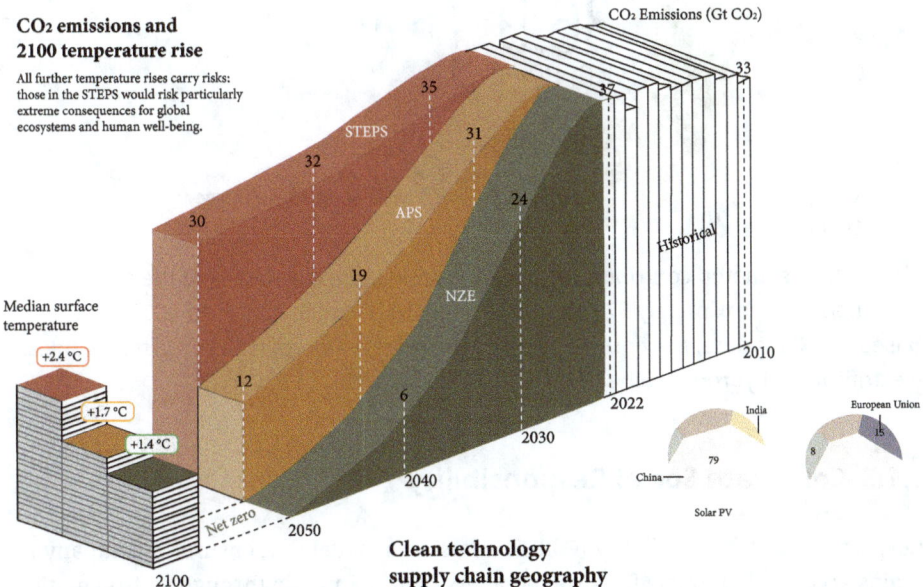

Figure 1.4 Projected trajectories of global greenhouse gas emissions and corresponding global temperature increases under three policy scenarios: STEP (stated policies), APS (announced pledges), and NZE (net zero emissions), based on IEA 2023 data [17]. The figure covers projections up to 2100 and highlights current emission levels (2022: 37 Gt CO_2)

A rapid energy transition is imperative to achieve short-, medium-, and long-term goals for reducing greenhouse gas emissions and realizing the NZE scenario. Energy efficiency and renewable energies are pivotal factors that promise substantial impact with low uncertainty [18]. According to IRENA figures [19], energy efficiency is projected to contribute 25% to the reduction of almost 37 Gt of annual CO_2 emissions from 2024 to 2050, as seen in Figure 1.5. This estimated percentage is equal to or higher than other strategies such as the generation and direct use of electricity based on renewable energies, the electrification of end-use sectors, the use of clean hydrogen and its derivatives, the bioenergy combined with carbon capture and storage, and the use of fossil fuel with carbon capture and storage. Hence, it is essential to implement energy efficiency programs, often comprising quite simple actions.

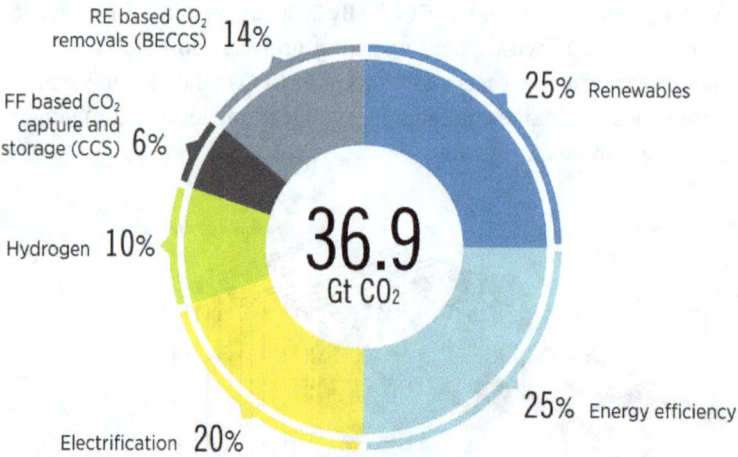

Figure 1.5 Estimated contributions of various mitigation strategies to the annual reduction of approximately 37 Gt of CO_2 emissions between 2024 and 2050 [19]. Based on IRENA data, the figure compares the roles of energy efficiency, renewables, electrification, hydrogen, bioenergy with CCS, and fossil fuels with CCS

1.1.3 Corporate Social Responsibility

Corporate social responsibility (CSR) is a business model that ensures a company remains accountable to itself, its stakeholders, and the public through self-regulation. By practicing CSR, companies acknowledge and take responsibility for their impact on economic, social, and environmental aspects [20]. Most companies incorporate varying degrees of CSR into their strategic planning, often with the key objective of reducing environmental impact. As mentioned, reducing greenhouse gas emissions is a primary societal concern and challenge. Therefore, companies typically measure this variable, with the carbon footprint being the most common metric, expressed in tonnes of CO_2-equivalent emissions.

Reducing the carbon footprint necessitates internal initiatives and collaborative efforts across the supply chain. As a result, it has become increasingly common for customers and giant corporations to require their suppliers – such as those in the plastics, components, packaging, and other product sectors – to establish and meet specific carbon footprint reduction targets and implement corresponding actions. Many countries are now implementing carbon taxes to incentivize further reductions in carbon footprints. Additionally, novel approaches such as carbon credits are gaining traction. Carbon credits enable companies to offset their greenhouse gas emissions by supporting activities that either reduce emissions (like renewable energy projects) [21, 22] or remove/sequester emissions (such as carbon capture and reforestation) [23, 24, 25].

Furthermore, there is a growing trend in the development of carbon footprint reduction labels. These labels are designed to communicate a product's efforts to reduce environmental impact directly to consumers, influencing their purchasing decisions based on environmental consciousness. Companies are driven to reduce their carbon footprint by combining factors including CSR commitments, supply chain pressures, tax incentives, and customer expectations. A primary strategy for achieving these goals is to reduce energy use, a significant contributor to carbon emissions. This approach yields tangible benefits, such as cost savings and tax advantages, and intangible benefits, such as enhanced reputation and alignment with evolving consumer preferences for sustainable products.

1.2 Strategies for Achieving Success in Enhancing Energy and Productivity Efficiency

Despite the recognized international economic and environmental benefits of energy efficiency, investments by small and medium enterprises (SMEs) often fall short of the achievable potential energy savings. This discrepancy is known as the "energy efficiency paradox", attributed to barriers and market failures that discourage investment in such programs [26]. A study of 280 SMEs in Europe revealed that only 10–25% of potential energy savings are realized [9]. Financial barriers include high investment costs, low capital availability in SMEs, and low profitability of investments. Additionally, the same study concluded that energy costs constitute a relatively small part of overall expenses for companies. SMEs face more demanding credit conditions than larger firms when seeking investment funding [9].

Other studies, such as the "Incombustion" program in Colombia, have focused on factors beyond the economic aspect [27]. Non-financial barriers identified in SMEs include limited internal expertise, lack of experience in identifying and implementing energy-saving projects, insufficient information and trust in available sources, challenges in accessing external expertise, and difficulty finding qualified consultants to support energy-saving initiatives. Furthermore, many SMEs do not use advanced metering technologies to monitor and control energy consumption effectively [28].

As a result of these challenges, energy audits often do not lead to high adoption rates of efficiency measures, typically remaining below 50%. These audits commonly focus on general components across industries like ventilation, lighting, motors, air compressors, and heating systems [29]. While addressing these areas does improve energy efficiency, it often overlooks specific and energy-intensive processes unique to each industry. This is particularly evident in polymer processing, where process-specific operations can account for 60–70% of total plant energy consumption [7], as shown in Figure 1.6. Optimizing these processes requires specialized knowledge of polymer processing techniques, materials, and technologies, which are not readily accessible.

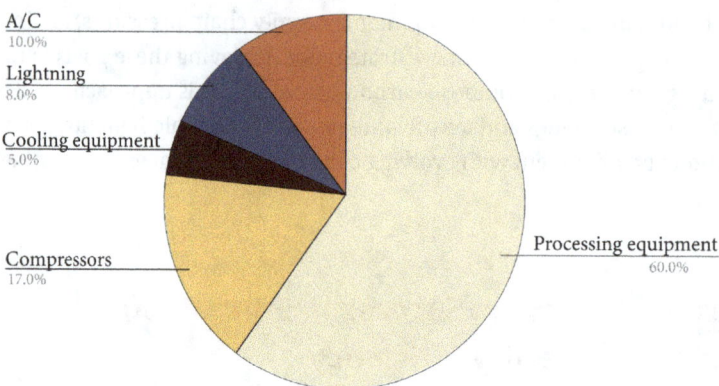

Figure 1.6 Typical distribution of energy consumption across different process stages in polymer processing plants, illustrating how specific operations, such as extrusion and molding, account for most energy use. Adapted from [7]

Four fundamental pillars are essential to succeed in energy management: a methodological approach, organizational structure, functional information systems, and organizational strategy and alignment. These pillars must be supported by a solid foundation: the staff's awareness, knowledge, and skills, as depicted in Figure 1.7 [30].

To strengthen the pillars and foundations supporting energy efficiency success, the following recommendations should be considered:

1. **Obtain top management commitment**: Top management commitment is the main element of organizational strategy. This commitment is evident when the organization adopts an efficient and rational energy use policy, allocates resources, and ensures policy compliance through monitoring.

2. **Evaluate the company's energy efficiency maturity and set development goals**: Under the methodological pillar, assessing the company's current maturity level is crucial for identifying gaps and developing progressive action plans. Various frameworks in the literature use "radar diagrams" to visualize current and desired states [7, 31], with evaluation criteria varying by framework. In the radar diagram of Figure 1.8, the evaluation is based on aspects detailed in Table 1.1.

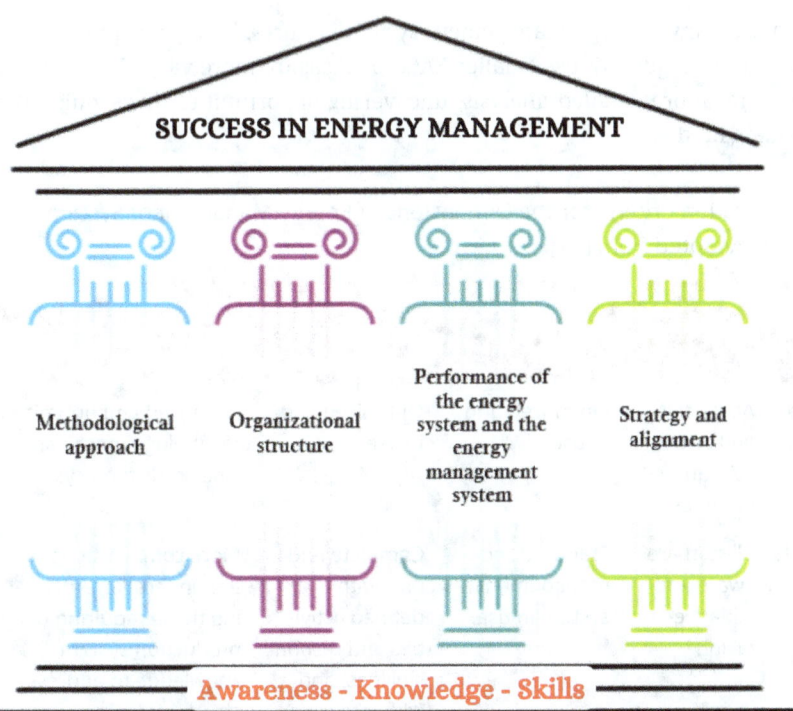

Figure 1.7 Conceptual framework illustrating the four main pillars – methodological approach, organizational structure, functional information systems, and strategy – and the foundational elements of staff awareness, knowledge, and skills that support successful energy management in companies [31]

3. **Select and plan energy accounting centers (EAC):** An EAC represents a selected unit for energy performance analysis [14] and allows its contribution to the company's energy costs to be determined. The EAC should meet some essential characteristics: it must be possible to individualize and totalize its consumption, all associated productivity and production information should be available, and it should be a significant part of the company's consumption and production. The EAC can be as large as the entire company or as small as a machinery component. This concept will be expanded in Chapter 3, where detailed information on the EAC is provided.

4. **Measure and store EAC data properly:** "What is not measured is not controlled". In the era of "Industry 4.0", real-time consumption and process speed monitoring have become accessible to companies of any size. The main barrier is the internal knowledge and skills needed to implement and leverage these technologies. Another option is the "software as a service" (SaaS) model, which has become extremely popular globally. SaaS allows specialized companies to set up an EAC monitoring instrument, and provide visualization, storage, and data analysis platforms for a monthly

fee. As a company's energy management system matures, it may be appropriate to measure more frequently, use smaller EACs, or measure more variables within the same EAC for more detailed analysis, uncovering opportunities that would otherwise remain hidden.

Table 1.1 Evaluation Criteria for the Dimensions of Energy Management System Development. Adapted from [31]

Maturity level	Awareness, knowledge, and skills	Methodological approach	Energy performance management	Implementation of best practices and available technologies
Optimized (80–100%)	Advanced and ongoing educational activities	Optimized and in use	Optimized and in use	Advanced and optimized technology for the specific production system
Integrated (60–80%)	All staff are aware and initiative-taking	Standardized management system in use	Complete and in use, with defined activities and responsibilities, and with intensive data collection and analysis activities having been developed	More complex actions are undertaken, including those requiring deep production system knowledge to address structural improvements
Systematic but not continuous (40–60%)	Management skills and awareness activities are being developed among all staff	Project management based on energy audit results	Still fragmented and poorly coordinated, but responsible parties are identified	Improvements are made only on commonly known elements, and investments are made if profitability is guaranteed
Occasional (20–40%)	Only basic technical knowledge exists	The use of validated methodologies is being debated	Occasional data collection	Improvements are made only to commonly known elements when they are low-cost and have a rapid return
Initial (0–20%)	Knowledge, awareness, and skills are fragmented among staff	Non-existent	Non-existent (based only on invoices)	Non-existent

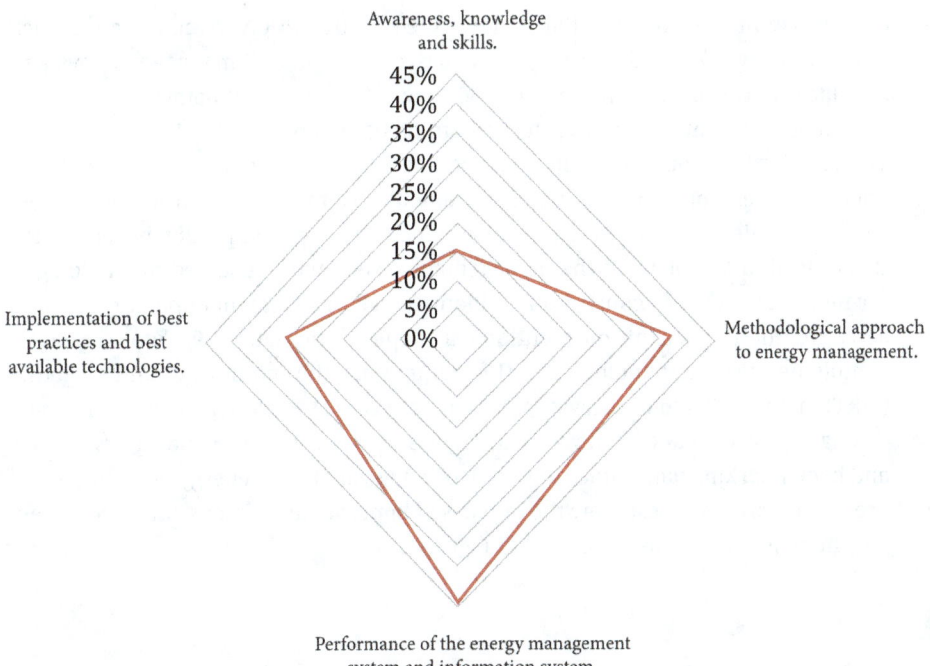

Figure 1.8 Radar chart comparing the current versus desired maturity levels of a company's energy management system, based on key evaluation dimensions such as knowledge, methodology, performance management, and technology implementation. Adapted from [31]

5. **Analyze available data**: Data analysis transforms raw data into actionable information essential for the continuous improvement of the EAC. When plotting EAC demand over time, many issues contributing to high energy consumption become immediately apparent without requiring elaborate analytics. For instance, it is possible to easily identify excessive consumption during non-production times, the largest energy consumers, equipment or processes failing to achieve stability, and irregular consumption patterns of equipment.

6. **Use validated methodologies with indicators and goals**: Controlling energy consumption in the industry involves implementing a methodology that ensures energy resources are used for maximizing economic benefit. This process, known as energy management, includes continuous energy consumption monitoring, planning, and implementing improvements to enhance energy efficiency and performance. It also includes incorporating basic elements tailored to each company's specific characteristics. An organized energy resource management system helps managers at all levels within an organization to achieve cost-effective energy use improvements, boost productivity, and improve product quality. Implementing an energy management

system within the company that follows the iterative PDCA (plan, do, check, act) strategy is advisable. This strategy is the basis for all the ISO management systems standards (ISO 9001, ISO 14001, ISO 25001, and ISO 50001). Whether formal or informal, it is recommended that this system be based on ISO 50001 standards [32]. This standard outlines administrative and documentary processes to ensure continuous improvement of the energy management system. However, it alone does not suffice to achieve objectives related to enhancing energy and productive efficiency. It is crucial to complement the management system with validated methodologies for monitoring energy consumption, identifying improvement opportunities, and assessing the impacts of interventions, as depicted in Figure 1.9. Two frequently complementary methodologies in this context are the monitoring and targeting (M&T) method [7] and the energy gap method (EGM) [14]. Equally essential is establishing key performance indicators (KPIs) or energy performance indicators (EnPIs), and benchmarking data to monitor, control, and evaluate the energy performance of production processes, estimate the impacts of improvement actions, and assess them post-implementation, as illustrated in Figure 1.10.

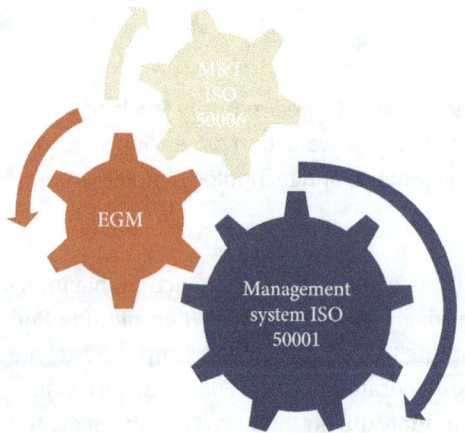

Figure 1.9

Framework for integrating key methodologies – monitoring and targeting (M&T) and the energy gap method (EGM) – to enhance energy performance and support continuous improvement in industrial energy management systems

7. **Integrate the operational fronts**: When the energy management system is perceived as disconnected from the company's priorities, achieving the expected results becomes challenging. Typically, companies operate on multiple fronts simultaneously. In a production plant, at least four distinct areas are usually identified: production, processes, quality, and technology management. Typically, each front operates with exclusive work teams and budgets, and establishes its management indicators, often operating independently. However, energy inefficiencies can arise in any of the mentioned areas, and so energy serves as a powerful integration force. Therefore, by using appropriate methodologies, energy management can unify the efforts across these various fronts within the industry. EGM [14] is among the methodologies capable of facilitating this integration, as will be detailed in Chapter 4.

Moreover, adopting an integrated management approach can optimize resources and streamline staff efforts in organizations with multiple management systems.

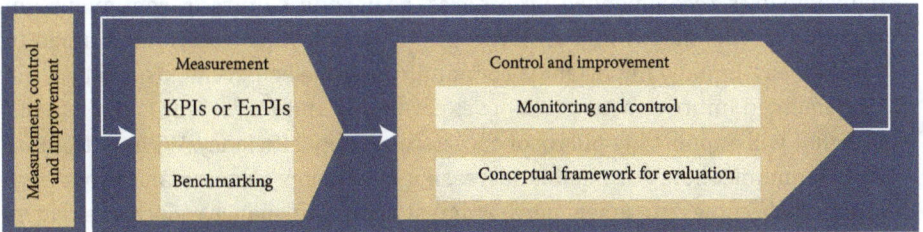

Figure 1.10 Visual representation of key performance indicators (KPIs) and benchmarking tools used to track, analyze, and improve energy performance in industrial processes, and so support continuous improvement and evidence-based decision-making

8. **Translate energy improvements into economic profits for the company**: Energy and productive efficiency represent one of any company's most substantial investment areas. Within a company, bridging the gap between technical staff overseeing the energy management system and decision-makers overseeing investments is crucial. Therefore, learning how to translate energy and productive efficiency enhancements into economic benefits is vital for estimating the internal rate of return (IRR) and payback period. It is estimated that 40% of improvements can be achieved without significant investments [7]. As improvements progress, investments become more substantial, making ensuring a return on investment increasingly challenging. Therefore, careful selection and prioritization of projects are necessary.

9. **Learn to identify improvement opportunities**: Identifying improvement opportunities for companies with immature energy management systems is typically straightforward. In polymer processing plants, there are well-known areas where significant gains in energy and productive efficiency can be achieved with minimal effort, low investment, and keen observational skills. Examples include replacing DC motors with AC motors, implementing VSDs on pumps and fans, addressing leaks in compressed air systems, insulating pipes and equipment to reduce heat loss or parasitic heat gains, and transitioning to LED lighting. Optimizing environmental conditions, adjusting processing parameters for lower energy consumption without compromising product quality, and preventing equipment overheating are effective strategies. Identifying additional opportunities becomes more challenging as efforts to enhance energy and productive efficiency progress. This requires higher technical expertise, particularly in plant operations and production processes. In such cases, it is essential to involve professionals with deep knowledge of polymer processing to build internal competencies and effectively drive continuous improvement initiatives.

10. **Train staff and communicate progress**: Addressing the growing challenges of energy and productive efficiency programs in polymer processing requires identifying and closing existing skill and knowledge gaps. The staff must be aligned with the company's recommended procedures, goals, and culture of continuous improvement. This means that all levels of the organization need to be trained to some extent. Equally important is communicating the achievements reached, the implemented improvements, and the expectations and plans. This is the foundation that will support the pillars of successful energy efficiency. Without training and communication, sustainably increasing a company's energy and productive efficiency becomes extremely hard. Progress and performance reports are crucial for communicating the company's energy management system's progress. Each report must specifically address elements related to the evolution of EnPIs, energy savings from implemented measures, emission reductions linked to these savings, and the calculation of ROI, among other metrics. It is recommended that at least four types of reports be prepared:

a. *Weekly or monthly report to detect energy waste*: These reports are essential for identifying areas of energy waste that need corrective actions. They feature tabulated results for monitoring EACs and often include graphical representations.

b. *Monthly or quarterly reports*: These reports inform the company about the progress in energy efficiency programs using cumulative sum (CUSUM) control charts, KPI evolution, and energy gap analysis for EACs. They include tabulated data, and typically feature a written summary that highlights progress made, discusses challenges encountered and goals achieved, forecasts energy consumption trends, and outlines actions being implemented to advance the program.

c. *Annual reports*: These comprehensive documents expand upon the results detailed in quarterly reports, providing an in-depth justification of overall outcomes. They focus on presenting and analyzing KPIs, assessing the system's achievements against annual plans, and outlining the energy management system's future trajectory for the upcoming year, including anticipated improvements and necessary investments.

d. *Diagnostic and evaluation reports*: These reports focus on each EAC. The diagnostic aims to understand the EAC's behavior or investigate unusual energy consumption patterns. To assess their impact, evaluations follow significant changes in technology, practices, or operating conditions affecting the EAC. They include graphical data, tables, and a section explaining findings, proposing solutions, and outlining improvement plans. Real-time monitoring graphs are particularly valuable in these evaluations.

Efficient and rational energy use and enhancing productive efficiency demands a comprehensive approach. While initial improvements are straightforward, sustaining and advancing these gains present more significant challenges. Efficient energy practices improve productivity, processes, quality, technology, and staff competencies. Companies can establish a robust foundation for an effective energy management system by methodically implementing these ten recommendations. Detailed discussions on these recommendations will be presented in later chapters of this book.

References

[1]   J. Bailey, "The steam age – Evolution of steam engines and the first steam locomotive", in: *Inventive Geniuses Who Changed the World*, Springer International Publishing, 2022, pp. 23–36, DOI: 10.1007/978-3-030-81381-9_3

[2]   B. D. Solomon, K. Krishna, "The coming sustainable energy transition: History, strategies, and outlook", *Energy Policy*, 2011, vol. 39, pp. 7422–7431, DOI: 10.1016/j.enpol.2011.09.009.

[3]   M. Economidou, V. Todeschi, P. Bertoldi, D. D'Agostino, P. Zangheri, L. Castellazzi, "Review of 50 years of EU energy efficiency policies for buildings", *Energy and Buildings*, 2020, vol. 225, article no. 110322, DOI: 10.1016/j.enbuild.2020.110322

[4]   E. Worrell, L. Bernstein, J. Roy, L. Price, J. Harnisch, "Industrial energy efficiency and climate change mitigation", *Energy Efficiency*, 2009, vol. 2, pp. 109–123, DOI: 10.1007/s12053-008-9032-8

[5]   "Annual Manufacturing Survey", Departamento Administrativo Nacional de Estadística (DANE), *https://www.dane.gov.co/index.php/en/statistics-by-topic-1/industry/annual-manufacturing-survey* [accessed 13 August 2024]

[6]   C. A. Vargas-Isaza, J. C. Posada-Correa, L. Y. Jaramillo-Zapata, L. A. García, "Consumos de energía en la industria del plástico: Revisión de estudios realizados", *Revista CEA*, 2015, vol. 1, p. 93, DOI: 10.22430/24223182.70

[7]   R. J. Kent, *Energy Management in Plastics Processing: Strategies, Targets, Techniques, and Tools* [4th edition], British Plastics Federation, 2024, *https://www.bpf.co.uk/Publications/energy-management-in-plastics-processing.aspx*

[8]   P. Montalbano, S. Nenci, D. Vurchio, "Energy efficiency and productivity: A worldwide firm-level analysis", *The Energy Journal*, 2022, vol. 43, pp. 93–116, DOI: 10.5547/01956574.43.5.pmon

[9]   J. Fresner, F. Morea, C. Krenn, J. Aranda Uson, F. Tomasi, "Energy efficiency in small and medium enterprises: Lessons learned from 280 energy audits across Europe", *Journal of Cleaner Production*, 2017, vol. 142, pp. 1650–1660, DOI: 10.1016/j.jclepro.2016.11.126

[10] "Climate Change 2007: Synthesis report", IPCC, 2008, *https://www.ipcc.ch/site/assets/uploads/2018/02/ar4_syr_full_report.pdf*

[11] "Energy Technology Perspectives 2017: Catalysing energy technology transformations", OECD, 2017, DOI: 10.1787/energy_tech-2017-en

[12] K. Onarheim, A. Arasto, "Market-driven future potential of Bio-CC(U)S" [Workshop summary], IEA Bioenergy, 2017, *https://task41project5.ieabioenergy.com/wp-content/uploads/sites/7/2017/10/Market-driven-future-potential-of-Bio-CCUS.pdf*

[13] A. B. Lovins, "Four revolutions in electric efficiency", *Contemporary Economic Policy*, 1990, vol. 8, pp. 122–141, DOI: *10.1111/j.1465-7287.1990.tb00649.x*

[14] O. Estrada, I. D. López, A. Hernández, J. C. Ortíz, "Energy gap method (EGM) to increase energy efficiency in industrial processes: Successful cases in polymer processing", *Journal of Cleaner Production*, 2018, vol. 176, pp. 7–25, DOI: 10.1016/j.jclepro.2017.12.009

[15] V. Masson-Delmotte et al. (eds.), "Global warming of 1.5°C: An IPCC special report on the impacts of global warming of 1.5°C above pre-industrial levels and related global greenhouse gas emission pathways, in the context of strengthening the global response to the threat of climate change, sustainable development, and efforts to eradicate poverty", Cambridge University Press, 2022

[16] "Energy Technology Perspectives 2015", OECD, 2015, DOI: 10.1787/energy_tech-2015-en

[17] "World Energy Outlook 2023", OECD, 2023, DOI: 10.1787/827374a6-en

[18] S. Sen, S. Ganguly, "Opportunities, barriers and issues with renewable energy development – A discussion", *Renewable and Sustainable Energy Reviews*, 2017, vol. 69, pp. 1170–1181, DOI: 10.1016/j.rser.2016.09.137

[19] "World Energy Transitions Outlook 2022: 1.5 °C pathway", International Renewable Energy Agency (IRENA), 2022, *https://www.irena.org/-/media/Files/IRENA/Agency/Publication/2022/Mar/IRENA_World_Energy_Transitions_Outlook_2022.pdf*

[20] M. Nurunnabi, J. Esquer, N. Munguia, D. Zepeda, R. Perez, L. Velazquez, "Reaching the sustainable development goals 2030: Energy efficiency as an approach to corporate social responsibility (CSR)", *GeoJournal*, 2020, vol. 85, pp. 363–374, DOI: 10.1007/s10708-018-09965-x

[21] M. Pérez De Arce, E. Sauma, J. Contreras, "Renewable energy policy performance in reducing CO_2 emissions", *Energy Economics*, 2016, vol. 54, pp. 272–280, DOI: 10.1016/j.eneco.2015.11.024

[22] Z. Yang, M. Zhang, L. Liu, D. Zhou, "Can renewable energy investment reduce carbon dioxide emissions? Evidence from scale and structure", *Energy Economics*, 2022, vol. 112, article no. 106181, DOI: 10.1016/j.eneco.2022.106181

[23] E. A. Duque-Grisales, J. A. Patiño-Murillo, J. Duque-Marín, S. Giraldo-Giraldo, J. A. Acosta-Strobel, "Can green bonds boost the development of energy projects in Colombia? An opportunity for responsible and sustainable investment", *International Journal for Management Science and Technology (IJMST)*, 2024, vol. 11, pp. 487–496, DOI: 10.15379/ijmst.v11i1.3685

[24] P. Ortiz-Grisales, J. Patiño-Murillo, E. Duque-Grisales, "Comparative study of computational models for reducing air pollution through the generation of negative ions", *Sustainability*, 2021, vol. 13, article no. 7197, DOI: 10.3390/su13137197

[25] *Negative Emissions Technologies and Reliable Sequestration: A Research Agenda*, National Academies Press, 2019, DOI: 10.17226/25259

[26] D. P. Van Soest, E. H. Bulte, "Does the energy-efficiency paradox exist? Technological progress and uncertainty", *Environmental and Resource Economics*, 2001, vol. 18, pp. 101–112, DOI: 10.1023/A:1011112406964

[27] R. Manrique, D. Vásquez, G. Vallejo, F. Chejne, A. A. Amell, B. Herrera, "Analysis of barriers to the implementation of energy efficiency actions in the production of ceramics in Colombia", *Energy*, 2018, vol. 143, pp. 575–584, DOI: 10.1016/j.energy.2017.11.023

[28] J. Leiva, A. Palacios, J. A. Aguado, "Smart metering trends, implications and necessities: A policy review", *Renewable and Sustainable Energy Reviews*, 2016, vol. 55, pp. 227–233, DOI: 10.1016/j.rser.2015.11.002

[29] K. Bunse, M. Vodicka, P. Schönsleben, M. Brülhart, F. O. Ernst, "Integrating energy efficiency performance in production management – Gap analysis between industrial needs and scientific literature", *Journal of Cleaner Production*, 2011, vol. 19, pp. 667–679, DOI: 10.1016/j.jclepro.2010.11.011

[30] V. Introna, V. Cesarotti, M. Benedetti, S. Biagiotti, R. Rotunno, "Energy management maturity model: An organizational tool to foster the continuous reduction of energy consumption in companies", *Journal of Cleaner Production*, 2014, vol. 83, pp. 108–117, DOI: 10.1016/j.jclepro.2014.07.001

[31] M. Benedetti et al., "Maturity-based approach for the improvement of energy efficiency in industrial compressed air production and use systems", *Energy*, 2019, vol. 186, article no. 115879, DOI: 10.1016/j.energy.2019.115879

[32] "ISO 50001: Developing an energy management system", ISO, 2018, *https://www.iso.org/iso-50001-energy-management.html*

2

Practical Concepts of Energy in Polymer Processing: A View from Macroscopic Energy Balances

Omar Estrada, Farid Chejne, Juan Carlos Maya

Mass and energy are two concepts intrinsically related in both thermodynamics and industry. From a thermodynamic perspective, an energy-efficient system is also one that optimally uses matter. In the polymer material transformation industry, raw materials (mass) constitute the primary cost, typically exceeding 50% of total expenses. At the same time, energy, usually categorized as an overhead, ranks as the second or third largest cost, depending on the region and the degree of industrial automation [1], as shown in Figure 2.1. This figure also highlights a common industry trend: more effort is often focused on optimizing areas with lower production costs.

A decisive and systematic approach to improving energy efficiency yields a significant side benefit – enhanced production efficiency. When energy efficiency increases, production costs decrease, leading to maximized profits and a stronger competitive position in the market. Additionally, improving energy efficiency has other far-reaching benefits, such as reducing the environmental impact of industrial processes [1] and enhancing working conditions and employee engagement [1]. For these reasons, monitoring the power consumption of production equipment and applying validated methodologies to drive continuous improvement is an industrial necessity. Before delving into the application of these methodologies in an industrial setting, it is crucial to first understand energy efficiency through fundamental thermodynamic principles.

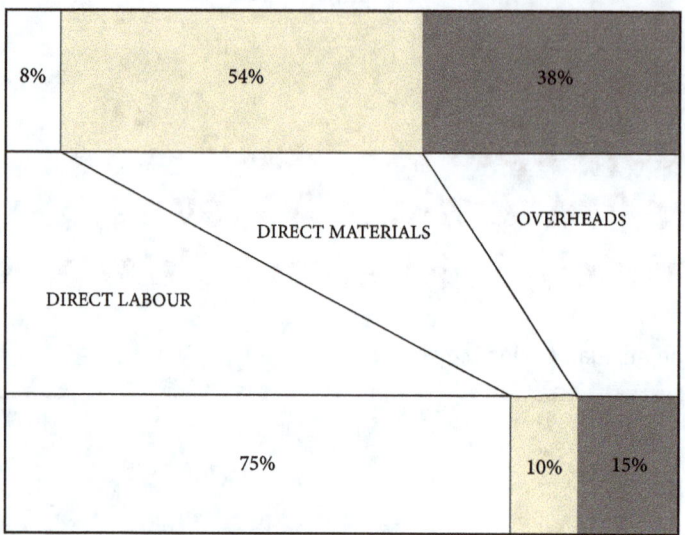

WHERE ARE THE EFFORTS?

Figure 2.1 Typical cost distribution in the polymer transformation industry [1]; The bottom of the figure shows the percentage of efforts typically devoted to improvement in the industry, and the top shows the approximate distribution of the concept to production costs

2.1 Energy and Power

Energy is the capacity of a system to induce transformations over itself or other systems through work or heat. Therefore, to achieve these transformations, the system consumes energy. Power refers to the rate of energy flow, or energy per unit of time, that the system requires to generate these transformations. In this context, the system demands power and consumes energy. The terms demand, power, and energy flow are used interchangeably, just as consumption and energy refer to the same concept.

In the International System of Units (SI), energy is measured in joules [J], while power is expressed as joules per second or watts [J/s or W]. Energy and energy flow or power are related by Equation 2.1.

$$E = \int_{t=t_1}^{t=t_2} \dot{E}(t) \cdot dt \qquad (2.1)$$

where:

E is energy [J]

$\dot{E}$ is energy flow [W]

t is time [s].

2.2 Energy Efficiency

Energy transfer and the capacity for transformation or energy content are not the same. All systems conduct energy transformations, and within these transformations, there is unavoidable irreversibility, especially when they involve industrial processes. The generation of irreversibility is the reason we talk about energy efficiency.

To understand energy efficiency, it is essential first to grasp how a system consumes energy or demands an energy flow. This requires performing a macroscopic energy balance over a "control volume". The control volume defines the boundaries of the system, as illustrated in Figure 2.2.

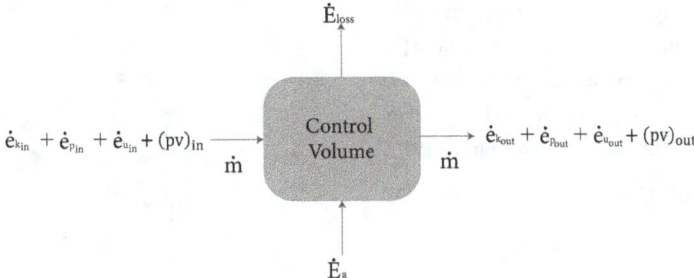

Figure 2.2 Energy balance in a controlled volume. The figure shows a control volume into which energy and mass enter, generating energy losses and a change in the kinetic, potential and internal energy that passes through the control volume

In the control volume shown in Figure 2.2, under steady-state or quasi-steady-state conditions, if energy enters the control volume or is lost to the surroundings, the mass passing through the control volume will gain or lose energy. This results in a change in its energy state between the inlet and outlet. The mass carries four types of energy: kinetic, potential, internal energy, and flow work. Energy balances can be conducted either in terms of total energy or energy flow; in this case, the balance will be performed using energy flow for a system in a stationary state, as represented in Equation 2.2 and Equation 2.3.

$$\dot{E}_a + \dot{e}_{k_{in}} + \dot{e}_{p_{in}} + \dot{e}_{u_{in}} + \dot{m}(pv)_{in} = \dot{E}_{loss} + \dot{e}_{k_{out}} + \dot{e}_{p_{out}} + \dot{e}_{u_{out}} + \dot{m}(pv)_{out} \qquad (2.2)$$

$$\dot{E}_a = \dot{E}_{loss} + \Delta\dot{e}_k + \Delta\dot{e}_p + \Delta\dot{e}_h \qquad (2.3)$$

where:

$$\Delta\dot{e}_u + \dot{m}\Delta(pv) = \Delta\dot{e}_h \qquad (2.4)$$

The flow of kinetic energy, potential energy, and internal energy is described by the following equations:

$$\dot{e}_k = \frac{1}{2}\dot{m}v^2 \tag{2.5}$$

$$\dot{e}_p = \dot{m}gH \tag{2.6}$$

$$\dot{e}_h = \dot{m}(h) \tag{2.7}$$

where:

$\dot{e}_k$ is the kinetic energy flow [W]

$\dot{e}_p$ is the potential energy flow [W]

$\dot{e}_u$ is the internal energy flow [W]

$\dot{e}_h$ is the change in internal energy flow plus the change in flow work [W]

$\Delta\dot{e}_k = \dot{e}_{k_{out}} - \dot{e}_{k_{in}}$ is the change in kinetic energy flow [W]

$\Delta\dot{e}_p = \dot{e}_{p_{out}} - \dot{e}_{p_{in}}$ is the change in potential energy flow [W]

$\Delta\dot{e}_u = \dot{e}_{u_{out}} - \dot{e}_{u_{in}}$ is the change in internal energy flow [W]

$\dot{E}_a$ is the energy flow added to the system [W]

$\dot{E}_{loss}$ is the loss of energy out of the control volume [W]

$\dot{m}$ is the mass flow rate [kg/s]

v is the velocity [m/s]

g is gravity [m/s^2]

H is the height [m]

h is the specific enthalpy [J/kg]

p is the pressure [Pa].

v_s is the specific volume [m^3/kg].

Analyzing Equation 2.3 reveals that, in the absence of energy losses ($\dot{E}_{loss} = 0$), all the energy supplied to the system is converted into transformations of equal magnitude in the matter flowing through the control volume, as represented by changes in its kinetic, potential, and/or internal energy. In this scenario, the system operates with 100% efficiency. Consequently, the magnitude of energy lost from the system is linked to its inefficiencies.

We can define energy efficiency (η_E) as follows:

$$\eta_E = 1 - \frac{\dot{E}_{loss}}{\dot{E}_a} = \frac{\Delta\dot{e}_k + \Delta\dot{e}_p + \Delta\dot{e}_h}{\dot{E}_a} \tag{2.8}$$

No system can achieve 100% efficiency, as each system is subject to specific thermodynamic constraints that establish a maximum achievable energy efficiency under ideal conditions. It is crucial to consider this maximum limit; for instance, an energy efficiency of 50% may be acceptable if the system's thermodynamic constraints only allow for a maximum of 60%. This topic will be further explored to provide a comprehensive definition of energy efficiency in industrial systems.

2.3 Energy Use in Polymer Processing

In the transformation processes of plastics and rubber, certain energy transformations are more significant than others, influenced by the properties of these materials and the characteristics of the processes involved. To illustrate this, we will analyze the example presented in Figure 2.3, which depicts an extruder operating with polypropylene homopolymer (PP-H).

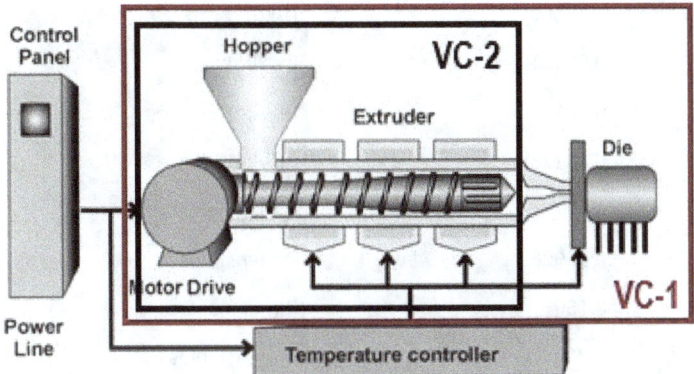

Figure 2.3 Definition of control volumes for an extrusion system: VC-1 is the control volume that encompasses both the plastification unit and the die; VC-2 is the control volume that includes only the plastification unit

The change in internal energy of the polymer is given by:

$$\Delta \dot{e}_u = \dot{m}(\Delta h - (\Delta p \cdot v_s + p \cdot \Delta v_s)) \tag{2.9}$$

Enthalpy change (Δh) represents the change in thermal energy per unit mass experienced by a moving fluid and is always significant in polymer processing. Strictly speaking, it reflects the energy that must be added or removed per unit of mass to achieve a temperature change. The term $\Delta p \cdot v_s$ denotes the gain or loss of mechanical energy per unit mass between the inlet and outlet of the control volume due to pressure changes generated in the system; this is also referred to as pumping power. The term $\Delta p \cdot v_s$ represents the gain or loss of energy per unit mass due to changes in density or specific volume caused by expansion or compression phenomena. In VC-1 of Figure 2.3, the inlet and outlet pressure are close to atmospheric pressure, making $\Delta p \cdot v_s$ and the $\Delta p \cdot v_s$ term negligible in magnitude compared to the change in enthalpy Δh (see Table 2.1). For $\Delta p \cdot v_s$ to be significant in polymer fluid processing, the pressure levels in the system must be exceedingly high, allowing these fluids to be treated as incompressible.

Now, if we conduct the same analysis using control volume VC-2 from Figure 2.3, the outlet pressure is no longer atmospheric, resulting in the $\Delta p \cdot v_s$ term exhibiting a significant magnitude that cannot be overlooked, as indicated in Table 2.1.

Table 2.1 Calculation of Internal Energy for Examples Illustrated in Figure 2.3

Parameter	Value	Units
$h_{23°C}$	39,081	[J/kg]
$h_{198°C}$	552,302	[J/kg]
$h_{192°C}$	534,560	[J/kg]
$v_{s(23°C,1bar)}$	0.00111	[m³/kg]
$v_{s(198°C,1bar)}$	0.00143	[m³/kg]
$v_{s(192°C,1bar)}$	0.00123	[m³/kg]
$\frac{\Delta p \cdot v_s}{\Delta h}$ (VC-1)	0	Dimensionless
$\frac{p \cdot \Delta v_s}{\Delta h}$ (VC-1)	0.000062	Dimensionless
$\frac{\Delta p \cdot v_s}{\Delta h}$ (VC-2)	0.150	Dimensionless
$\frac{p \cdot \Delta v_s}{\Delta h}$ (VC-2)	0.015	Dimensionless
Data extracted from [2]		

According to the example presented in Figure 2.3, if there are no pressure changes between the inlet and outlet of the control volume:

$$\Delta \dot{e}_u = \dot{m}\Delta h \tag{2.10}$$

On the other hand, if there are pressure differences between the inlet and outlet of the defined control volume:

$$\Delta \dot{e}_u = \dot{m}(\Delta h - \Delta p \cdot v_s) \tag{2.11}$$

The change in internal energy is several orders of magnitude higher than the change in kinetic and potential energy, as shown in Table 2.2. This indicates that, in polymer processing, most of the energy supplied is consumed to achieve the change in the internal energy of the material. Specifically, this energy is primarily consumed to promote enthalpy changes through heating and cooling processes, which are essential for plastifying and shaping the products.

Table 2.2 Macroscopic Energy Calculations of VC-1 in Figure 2.3

Parameter	Value	Units
$\dfrac{\Delta\dot{e}_p}{\Delta\dot{e}_u} = \dfrac{g \cdot \Delta H}{\Delta h}$	0.000363	Dimensionless
$\dfrac{\Delta\dot{e}_k}{\Delta\dot{e}_u} = \dfrac{1}{2}\dfrac{\Delta(v^2)}{\Delta h}$	0.0000974	Dimensionless

In conclusion, in polymer processing, from a thermodynamic perspective, the energy changes in the processes primarily promote the change in the internal energy of the polymer, since in these processes:

$$\frac{\Delta\dot{e}_k}{\Delta\dot{e}_u} \ll 1; \frac{\Delta\dot{e}_p}{\Delta\dot{e}_u} \ll 1 \tag{2.12}$$

Therefore, energy efficiency in polymer processing can be simplified and defined as:

$$\eta_E = \frac{\Delta\dot{e}_u}{\dot{E}_a} = \frac{\dot{m}\,(\Delta h - \Delta p \cdot v_s)}{\dot{E}_a} \tag{2.13}$$

The specific volume and enthalpy data have the form presented in Figure 2.4 and Figure 2.5 respectively, for the case of polypropylene. Information about other polymers and other properties such as thermal conductivity and heat capacity can be found in the literature [3].

The behavior of specific enthalpy changes according to the type of polymer. For amorphous polymers, it may be calculated from Equation 2.14:

$$\Delta h = C_p \Delta T \tag{2.14}$$

while in semi-crystalline polymers it is given by Equation 2.15:

$$\Delta h = C_p \Delta T + \Delta H_f \tag{2.15}$$

where:

C_p is the heat capacity

T is the temperature

ΔH_f is the crystal melting enthalpy.

In Figure 2.4, atactic polypropylene is amorphous, and isotactic polypropylene (which is the majority of commercial PP) is semicrystalline. So, the former has a linear behavior and the latter does not.

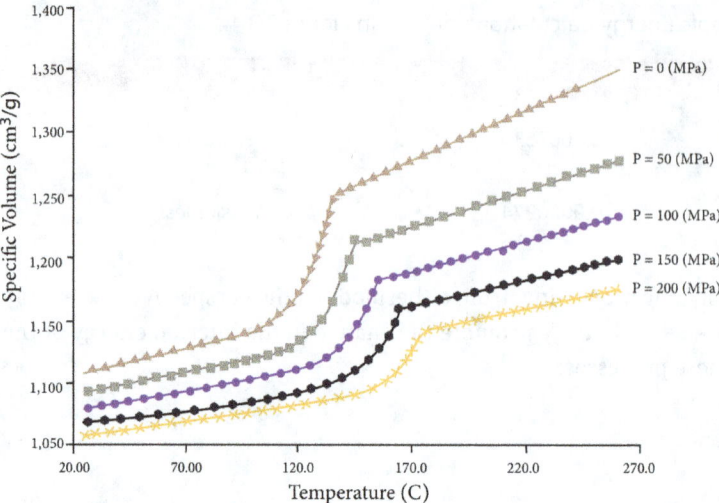

Figure 2.4 PvT diagram for polypropylene; The PvT diagram presents the behavior of the specific volume of a substance as a function of its temperature and the pressure exerted on it

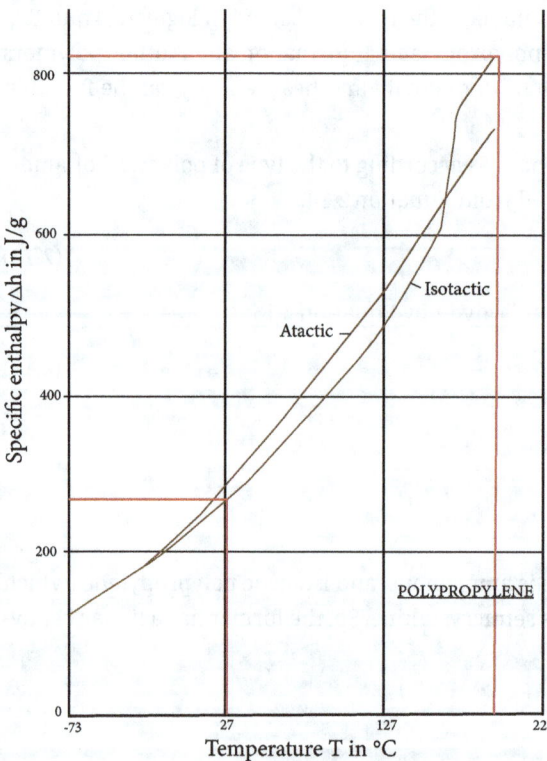

Figure 2.5
Specific enthalpy vs. temperature diagram for polypropylene; The graph shows the specific enthalpy for isotactic and atactic polypropylene, which represents the energy required to be supplied to one kg of the material to go from the reference temperature to the temperature selected in the diagram

2.4 Specific Energy Consumption and Energy Efficiency

In the industrial sector, assessing the energy consumed in producing products that meet market standards and satisfy customer quality requirements is crucial. The indicator that provides this information is the "net specific energy consumption" or SEC_n. This book uses the term net specific energy consumption because multiple SEC values will be defined, while in other publications, it is often referred to simply as specific energy consumption (SEC). It is defined as follows:

$$SEC_n = \frac{E_a}{W_c} \tag{2.16}$$

where:

SEC_n is the net specific energy consumption [kWh/kg]

E_a is the energy supplied to the control volume [kWh]

W_c is the compliant production in the control volume [kg]. Compliant production refers to a manufacturing process in which finished products meet all quality requirements, technical specifications, and applicable regulations.

If the cost of energy is known, the energy cost for obtaining compliant products can be estimated as follows:

$$ECP = SEC_n \cdot EC \tag{2.17}$$

where:

ECP is the energy cost of the product [US\$/kg]

EC is the cost of energy [US\$/kWh].

The SEC_n, when effectively managed, serves as a valuable measure of performance and efficiency. However, if mismanaged, it can lead to misleading conclusions. Generally, the lower the SEC_n, the better the energy efficiency of the production process, and in some cases, the better the energy performance. However, a reduction in this indicator does not always reflect effective actions taken to enhance the energy performance of the processes. While increasing energy efficiency is beneficial, improving energy performance is even more important. This distinction between efficiency and performance will be discussed further.

The question that arises is how much SEC_n can be reduced. The theoretical minimum value is reached in a 100% efficient system, where all the energy is used solely to increase the internal energy of the polymer, Δe_u. From this concept, we can define a new indicator, the "thermodynamic minimum specific energy consumption" or SEC_t, which is expressed by the following equation:

$$SEC_t = \frac{\Delta e_u}{W_t} \tag{2.18}$$

where:
W_t is the total production, including both compliant and non-compliant products. Usually, the units of internal energy are joules [J]. This unit is equivalent to kWh, where 1 kWh = 3,600,000 J.

The energy efficiency of the process can be defined in terms of specific energy consumption such as:

$$\eta_E = \frac{SEC_t}{SEC_n} \cdot \frac{W_t}{W_c} \tag{2.19}$$

The difference $SEC_n - SEC_t$ represents the energy gap between the actual performance of the control volume and the ideal performance in a 100% efficient system. This gap highlights the inefficiencies within the control volume, including the energy consumed in producing non-compliant products, and reflects the total energy lost during the process. It is important to understand the difference between wasted energy and lost energy.

Waste energy is generated in all energy processes. This waste energy may be available and reusable energy, or it may be scarcely available and unusable. Both the unavailable energy and the available energy that is not reused become energy losses. For example, cooling air in a pipeline is heated. That heat is available and can be used to heat the material at the extruder inlet. If it is not used, that energy is lost. Therefore, you must ask yourself:

- Can the discarded energy be recovered or reused?

- Is the technology suitable for maximizing energy efficiency?

- Are the conditions appropriate for technology to achieve the best possible energy transformation?

- Can the processes or technologies be optimized, and what would be the associated costs?

- What is the environmental impact of the energy resource being used?

The following sections introduce methodologies designed to help answer many of these questions. This book focuses on two key approaches: the "monitoring and targeting" (M&T) methodology, used to control and optimize energy consumption and performance, and the "energy gap method" (EGM), a diagnostic tool for identifying energy inefficiencies.

References

[1] R. J. Kent, *Energy Management in Plastics Processing: Strategies, Targets, Techniques, and Tools* [4th edition], British Plastics Federation, 2024, *https://www.bpf.co.uk/Publications/energy-management-in-plastics-processing.aspx*

[2] C. A. Naranjo, E. M. del Pilar Noriega, M. J. D. Sierra, J. R. Sanz, *Injection Molding Processing Data* [2nd edition] ("Plastics Pocket Power" series), Hanser Publications, 2019

[3] B. Weidenfeller, M. Höfer, F. R. Schilling, "Thermal conductivity, thermal diffusivity, and specific heat capacity of particle filled polypropylene", *Composites Part A: Applied Science and Manufacturing*, 2004, vol. 35, pp. 423–429, DOI: 10.1016/j.compositesa.2003.11.005

3
Monitoring and Targeting (M&T)

Julián Patiño, Omar Estrada, Farid Chejne

The "monitoring and targeting" (M&T) methodology is a widely used and disseminated tool to support energy efficiency and performance improvement processes in virtually any area where energy is consumed, such as industry. M&T was developed in the 1980s in the United Kingdom and quickly became popular, as it proved to be a methodology that can achieve significant impacts in reducing energy consumption, energy costs, and the environmental impact of consumption in the generation of greenhouse gases [1]. Dr. Robin Kent, in his book *Energy Management in Plastics Processing: Strategies, Targets, Techniques, and Tools* [2], where he compiles his experience of more than 500 polymer processing companies around the world, points out that it is possible to reduce energy consumption in many plants by approximately 20%, either without the need to make investments, or through investments that are recovered in less than 9 months. 50% of these savings can be achieved through M&T, improving process management, and establishing relevant management control systems.

M&T has two crucial components: monitoring, whose essential objective is establishing the existing pattern of energy consumption, and targeting, which seeks to identify the level of energy consumption that is desirable as a management objective according to the patterns identified during monitoring. Another fundamental stage is reporting, which makes it possible to socialize energy management at all levels of the organization through the control of management indicators and actions for fulfilling improvement objectives, so that the results can be followed up and the impacts of the initiatives undertaken can be demonstrated. For this reason, some authors refer to the method as "monitoring, targeting, and reporting" [3].

The first step in adequately applying the M&T method at the industrial level is defining the production unit subjected to analysis using the methodology. This unit is called the energy accounting center (EAC). The "energy gaps" method also uses this concept, which will be presented later.

3.1 Energy Accounting Center

An EAC is a specific unit within the polymer processing plant where detailed energy consumption monitoring and analysis is performed [4]. These centers can be as large as an entire plant or as small as a single piece of equipment or process [2], as shown in Figure 3.1.

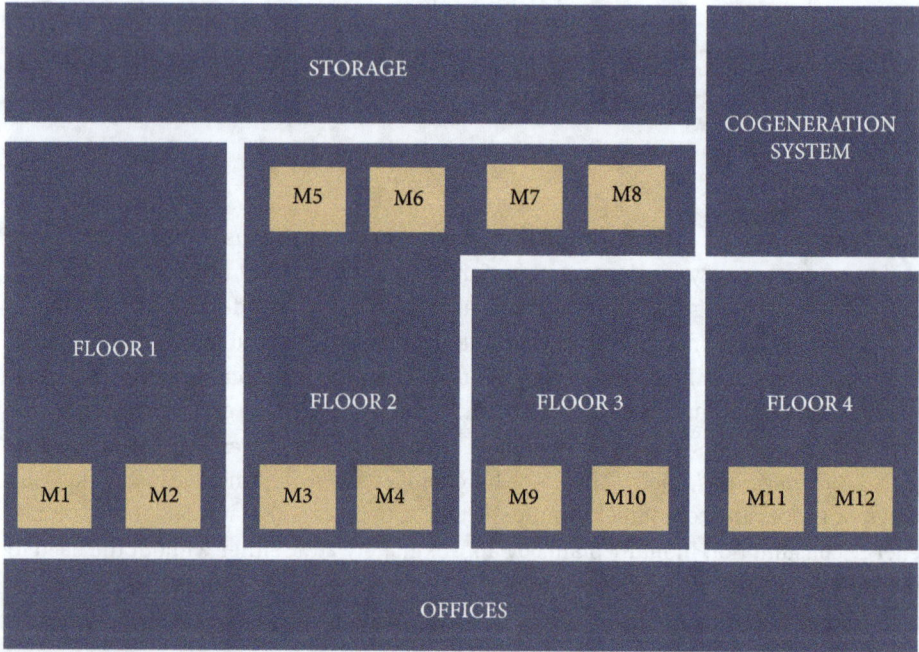

Figure 3.1 Illustration of possible configurations for selecting EACs in a polymer processing facility. The figure shows how EACs can be defined at different aggregation levels – from the entire plant to individual equipment – for effective monitoring and control of energy consumption

Assertive identification of EACs allows companies to focus their monitoring, optimization, and management efforts on the areas representing the highest energy costs [5]. By segmenting energy consumption into manageable units, EACs facilitate the detection of specific inefficiencies and the implementation of targeted improvements, ensuring that each part of the process contributes to reducing energy consumption and improving productivity. The EAC can comprise one or more sections or even the entire enterprise, just taking the first steps toward energy management. As the energy management system evolves, granular targeting is essential to achieve effective and sustainable energy management over the long term [2].

The EAC has inputs and outputs that must correctly identify and generate specific information requirements. A good way to identify the correct inputs and outputs is through the "black box" analogy. It can be said that energy accounting is a black box placed over the unit on which the energy performance analysis is to be performed [6]. Thus, the inputs and outputs of the EAC are schematized as in Figure 3.2 [7]. Raw materials and energy enter an EAC, and the EAC outputs information on compliant production, non-compliant production, and energy losses.

The M&T method requires at least the following information to be very clearly defined:

- The energy consumption of the EAC in the period of analysis

- Compliant production associated with the EAC within the same period.

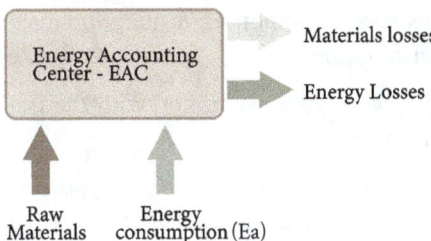

Compliant
product (Wc)

Figure 3.2

Schematic representation of the EAC as a control volume. The diagram identifies key inputs (energy and raw materials) and outputs (compliant products, non-compliant products, and energy losses), highlighting the fundamental structure for applying the M&T methodology

To comply with the first information requirement, the EAC must have a totalizer that ensures that it is possible to determine all its consumption without the interference of other consumers external to the EAC that may distort the results. The company's electrical network must be thoroughly reviewed and adjusted to achieve this. To comply with the second requirement, a sufficiently robust information system must be in place to associate production with energy consumption.

One of the main mistakes in analyzing an EAC's energy performance is to consider information about what happens within the black box. For example, if the EAC is defined as a flexible packaging production plant, all energy consumers inside the black box (processing equipment, air conditioners, lighting, general services, among others) should be associated with the compliant production that reaches the dispatch area. If 1 kg of compliant product went through extrusion, printing, and sealing, it is not 3 kg of compliant production because it went under the black box. Only 1 kg arrived as a finished product at the shipping area. All in-process inventory does not count. If the EAC is a hollow body blow-molding line in a manufacturing cell with its mill grinding burrs and non-compliant bottles for reintroduction into the process, the reprocessed material is not part of the non-compliant product because it is not leaving the EAC. If waste is generated during the analysis period and leaves the manufacturing cell, it is considered non-conforming production and is part of the material losses.

The cost center's energy consumption is characterized by the appropriate variables identified and measured. However, it is not the only necessary information, since the company must be able to determine the causes that generate the identified behavior to eliminate them when the energy performance of the process is deficient, or promote them when the performance improves. Therefore, it is very valuable to have data to establish quality problems, downtime distributions, process speed monitoring, and any other helpful information to make decisions to improve the performance of each EAC. Thus, EACs are not abstract entities. They are physical entities that are created with the following characteristics:

- An EAC is strategically selected for the focusing of improvement efforts

- A budget must be assigned to the EAC

- A person responsible for monitoring the EAC's performance must be assigned

- The electrical system must be adjusted, if necessary so that the consumption measurements correctly represent the EAC's consumption

- The process must be instrumented to monitor energy consumption with the frequency required by the type of analysis to be performed

- An information system must be implemented to provide production and process data that allows correlation of consumption with EAC behavior. The information system must meet the analysis requirements. If the monitoring of the EAC is planned daily, weekly, monthly, by production order, or in any other way, both the consumption monitoring and the information system must be congruent with such planning

- Production monitoring is not strictly necessary when a properly planned and executed information system is in place. However, some production data can be useful for correlating with energy consumption and monitoring the proper functioning of the information system. For example, tracking parts produced per minute in batch processes or meters per minute in continuous processes

- Relevant energy performance indicators (EnPIs) should be generated for each EAC

- Internal procedures should be formalized, so that each of the actors in the system understands their responsibilities and actions concerning the EAC

- Data analytics should be generated to convert measurements and records into relevant information for improvement

- Appropriate reports should be designed for each EAC to socialize the evolution of the EnPIs and the savings achieved in consumption and greenhouse gas (GHG) emissions. In addition, the financial results of the energy efficiency improvement actions carried out on the EAC must be shown, which are usually presented in terms of return on investment (ROI).

The EAC is analogous to the control volume defined in Chapter 2 insofar as it is the control volume over which energy management is performed. It is from the definition of the EAC in Figure 3.2 that the expression for the calculation of SEC_n (net specific energy consumption) of Equation 3.1 makes sense. This equation states that:

$$SEC_n \left[\frac{kWh}{kg} \right] = \frac{Energy_{consumption}\,[kWh]}{Production_{compliant}\,[kg]} = \frac{E_a}{W_c} \tag{3.1}$$

Based on the elements discussed so far, SEC_n is the ratio of EAC energy consumption (E_a) to compliant production (W_c) during the same period. The first important observation is that the compliant production of EAC at the industrial level can be measured in units, length, or mass according to how the user considers that the quantities produced or sold are best represented. However, the results are only comparable if they are in the same units of measurement. This is why compliant production is presented in units of mass (specifically in kilograms) in this book, and for all the examples that will be explained later. Furthermore, for reference purposes, it is much easier to find SEC_n values in units of kWh/kg than in other units.

3.2 Tools of the M&T Method Suitable for Data Analytics

The M&T method is very popular because it relies on tools that are very useful for tracking the energy efficiency and energy performance of an EAC. Among these tools are the energy consumption formula (ECF), which most of the time can be as simple as a linear regression represented by:

- The performance characteristic line (PCL)

- The activity based target (ABT), used to set improvement targets

- CUSUM diagrams, to make comparisons and evaluate the improvements expected or obtained with the decisions made to achieve the ABT.

3.2.1 Energy Consumption Formula (ECF)

As mentioned above, data about how the EAC consumes energy and generates output must be collected. The next stage of the monitoring process is to analyze the data to understand the dependence of energy consumption on production variables. This is achieved through an energy consumption formula (ECF) [8], from which the factors that drive EAC consumption could also be determined. The model used as ECF does not have to be complex, although it can be. From the M&T point of view, the only requirement is that the model produces a numerical value representing the expected energy consumption. It is essential to understand that there will be a different model for each energy source in the EAC: one for steam, one for electricity, one for fuels, and so on, as appropriate.

The production variables are associated with the inputs and outputs of the EAC. The factors that drive consumption are those determinants that cause significant changes in the intensity with which the EAC consumes energy. Some of these factors are the composition of production, the materials that are processed, the type of transformation processes used in the EAC and the distribution of their use, the technologies used by the EAC and the distribution of their use, among others. An example of this is a plant that has extrusion, injection, and blow-molding processes, and only one EAC is taken: the company as a whole. The intensity with which each of these processes consumes energy is different. As long as the production composition does not involve significant changes in the compliant production percentage of each process, the energy consumption intensity remains approximately constant. However, the intensity changes if there are substantial changes in the composition. When consumption intensity depends on a single factor or several factors that do not change substantially over time, energy consumption can be modeled as a linear relationship with a single variable. When it does not, the model will depend on several variables, and multivariate analysis is required.

3.2.1.1 Performance Characteristic Line (PCL)

There are different methods for analyzing energy consumption to determine potential savings. Regression analysis is the most used method for modeling energy consumption against a variable to determine the performance characteristic line (PCL). This technique is helpful for EAC when a single determinant driver predominantly influences consumption. In such circumstances, a linear relationship is good enough as a model. If there is more than a single energy source or a single driving factor, each process will have a different PCL, as processes are not equally demanding in terms of energy.

For polymer processing plants, a linear correlation between energy consumption and a single driving factor is acceptable when the coefficient of determination (R^2) is greater than or equal to 0.7 [2]. For an industrial plant EAC, compliant production is the variable that correlates with energy consumption when the data are measured in the same seasonal period or under the same criteria, as shown in Figure 3.3.

Figure 3.3 shows the determination of the PCL of an EAC for a recycled plastic agglomerated posts production plant. The energy consumption and the monthly compliant production of 2023 were plotted for its construction. This shows an adequate correlation with a straight line as it has a coefficient of determination (R^2) of 0.8718. This means that the dependence of the consumption (Energy$_{consumption}$) with the monthly compliant production (Production$_{compliant}$) of this plant can be modeled using the following equation:

$$\text{Energy}_{consumption}\,[\text{kWh}] = a\left[\frac{\text{kWh}}{\text{kg}}\right]\cdot\text{Production}_{compliant}\,[\text{kg}]$$
$$+b\,[\text{kWh}] \tag{3.2}$$

$$\text{Energy}_{consumption}\,[\text{kWh}] = 0.359\left[\frac{\text{kWh}}{\text{kg}}\right]\cdot\text{Production}_{compliant}\,[\text{kg}]$$
$$+1293.8\,[\text{kWh}] \tag{3.3}$$

where:

a is the slope of the straight line, also known as the variable consumption, and has units of kWh/kg. In this case, $a = 0.359$ kWh/kg. Variable consumption is associated with the cost of production, which can be reduced, but not eliminated, through improved efficiency

b is the intercept of the straight line, also known as the fixed consumption or base load, and has units of kWh. In this case, $b = 1293.8$ kWh. Fixed consumption is associated with expenses. It is wasted energy that is not directly associated with production. This energy consumption is unavoidable but reducible.

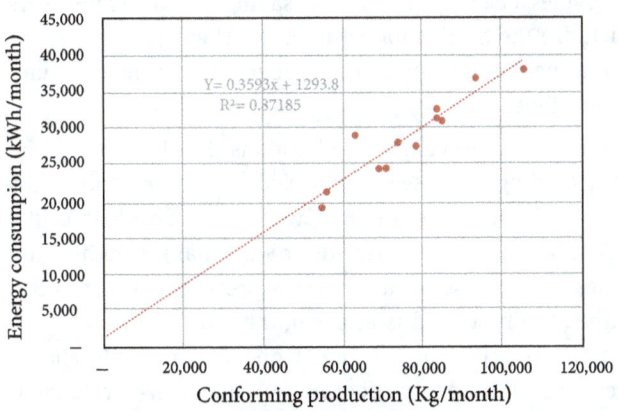

Figure 3.3 PCL for a plant producing recycled plastic agglomerate posts and planks. The plot shows a strong linear correlation ($R^2 = 0.8718$) between monthly energy consumption and compliant production during 2023, supporting use of a linear regression model for performance analysis

The base load is the energy consumed if there is no compliant production in an industrial plant. Of course, with zero compliant production in a factory, there is expected to be no energy consumption. However, it is common, and even inevitable, that moments of zero-compliant production occur, where a certain amount of energy is needed to keep the factory running (analogous to a car consuming energy while waiting at a red light). The quantification of these consumptions in the EAC becomes the fixed consumption or base load, and the plant's efforts should focus on minimizing the factors that make this value grow. These factors are:

- Production of non-compliant products

- Downtime

- Improper equipment startup, shutdown, and standby practices

- Consumption of equipment or components that do not need to be consumed to maintain availability

- The presence of oversized capacity consumers for the operation of the EAC

- The presence of consumers of low energy efficiency or high consumption necessary for the EAC but whose consumption does not depend on production (such as lights, air conditioners, surveillance, and monitoring systems, among others).

On the other hand, the variable load represents the minimum energy consumed by the cost center to produce each kilogram of compliant product when there are no energy losses for any reason. In this case, it is like a car's energy consumption per kilometer traveling at a constant speed. An EAC with a lower variable load indicates a more efficient EAC. Variable load also provides a basis for calculating savings related to the users of variable energy loads, which are the production equipment. These savings can be obtained by implementing operating, maintenance, best practices or technology management measures that increase efficiency.

Two critical observations arise from the previous considerations. The first is that the base load must be interpreted according to the seasonal period or criterion chosen for the graphical representation of the information. In other words, the base load of the analysis of the same EAC in two different seasonal periods or with changes in the criteria may assume different values. In the case of the example presented in Figure 3.3, since the information is monthly, the base load is also monthly and is equivalent to 1,294 kWh/month. The second is that the efficiency of the EAC does not change by changing the seasonal period or the criteria for the energy data analysis and, therefore, should remain approximately constant.

On the other hand, it is helpful to compare the energy performance information of the EAC with information documented for other similar EACs. This establishes whether the fixed and variable load values are reasonable or merit intervention. Table 3.1 and Table 3.2 present typical fixed load and variable load values based on industry data for different types of plastics production plants [2]. According to the same source and the experience of that author in polymer processing plants, the base load is usually 20–40% of the average total load. However, the same author indicates that any baseload percentage above 30% of the average total load generally warrants review and reduction. It is worth clarifying that the information in Table 3.1 corresponds to the complete plant, including the consumption of the general services of heating, cooling, compressed air, lighting, and other resources required for the operation of the plant. Therefore, the values in Table 3.1 do not apply when the EAC represents exclusively a production line and not the entire plant.

Table 3.1 Typical Site Base and Process Load Values Based on Industry Data for Plastic Production Processes [2] [a]

Process/site load	Base load [% of average total load]	Process load [kWh/kg]
Injection molding	30	0.9–1.6
Extrusion	31	0.4–0.7
Extrusion blow-molding	25	0.8–1.3
Rotational molding	38 (electricity) 12 (gas)	0.3–0.6 (electrical) 1.8–2.7 (gas)
Injection blow (injection + blowing steps)	—	1.0–1.6
Thermoforming (sheet forming step only)	—	0.1–0.3
Thermoforming (including extrusion)	—	0.5–1.2

[a] The average plant SEC is not useful for external benchmarking because the overall production rate is important, and it is essential that the benchmarking is adjusted for this production rate dependence. Now, the average plant SEC [kWh/kg] is always higher than the process load [kWh/kg] because the base load is included in the plant SEC, while the process load only considers the processing energy use.

Table 3.2 Average site SEC based on industry data for plastic production processes [a]

Process	Average site SEC [kWh/kg]	
	EURecipe [a]	Tangram [b]
Injection molding	3.118	3.138
Extrusion	1.506	1.316
Extrusion blow-molding	N/A	2.07
Rotational molding	5.828	4.85 (all fuel)

[a] Data from the EURecipe report (a survey of European plastics processors in 2005).

Continuing with the example presented in Figure 3.3, the average monthly consumption was 28,736 kWh/month, the base load determined through the PCL is 1,293.8 kWh/month, and the variable load is 0.359 kWh/kg. This is a base load percentage of 4.5%. Producing stakes or poles with bonded plastic resembles a thermoplastic extrusion process. The plant in question, when compared to similar plants (see Table 3.1), has a higher efficiency by having a lower variable load than expected for an extrusion plant

(0.359 kWh/kg vs. 0.4–0.7 kWh/kg) and a very low base load compared to the average consumption of this type of plant (4.5% vs. 31%). The plant does not have compressors or chillers among the general services, as cooling is done with rainwater and a cooling tower. It also indicates that the plant has good management practices for aspects that limit non-productive times, such as production scheduling, startup, shutdown, change of reference, and maintenance.

Monitoring the data quality is as essential as the correlation of the information. The higher the correlation coefficient, the better the PCL will represent the data. As previously stated, when energy data does not correlate properly, it may be an EAC with several driving factors. If this is not the case, the first thing to check is the congruence of the data. This involves the design and operation of the information system. The information system goes beyond the tools used to record information; the procedures establish what data, who, when, how, and where it is collected and recorded. One other issue is poor management of the process. If the R^2 value is low, the first thing to do is to check the data (collection methods, times, etc.). If the data is of high quality, then the central issue is in the management of the process and inconsistent management processes. This will also show poor quality, maintenance, and generally poor operational management.

If the data are checked and found to be congruent, there may be problems in how the data are used. When seasonal analysis periods are taken, during which there are changes in the driving factors, the correlation coefficient is reduced. This occurs when in the EAC, for example, equipment is changed, high-impact maintenance is performed, such as the recovery of plasticizing units, when products of very different energy intensity than usual enter the production composition, substantial changes are generated in the procedures for the operation of the equipment, new equipment is removed or enters the EAC, among others. In these cases, dividing the analysis into stages before and after the changes may be sufficient. When all previous elements have been ruled out as a potential cause of low coefficients of determination, the representation of energy consumption by alternative models should be explored. It should be noted that multifactor or nonlinear models will sometimes be more appropriate. As described below, mathematical models of greater complexity could be employed when circumstances require them and resources permit them.

3.2.1.2 Mixed Processes or Processes with Multiple Driving Factors

There are many cases where two or more factors determine the level of demand. There may be multiple drivers for energy use when the plant has a combination of processes, or the same energy source is used for different purposes whose consumption is not measured separately [2]. Typical examples are:

- Plants where two processes with very different energy intensities are carried out. For example, a plant where injection molding and extrusion are performed. Extrusion is much less energy-intensive than injection molding, and energy use will vary according to the relative volumes of material processed

- Plants where total energy consumption is analyzed and electricity is used not only for production but also to power critical additional process loads, such as casting, air conditioning, or heating. Part of the energy use will be due to the process (driving factor associated with production) and additional consumption (driving factor related to operating conditions)

- Plants where gas is used for production processes and heating. Part of the energy use will be due to the production process and space heating.

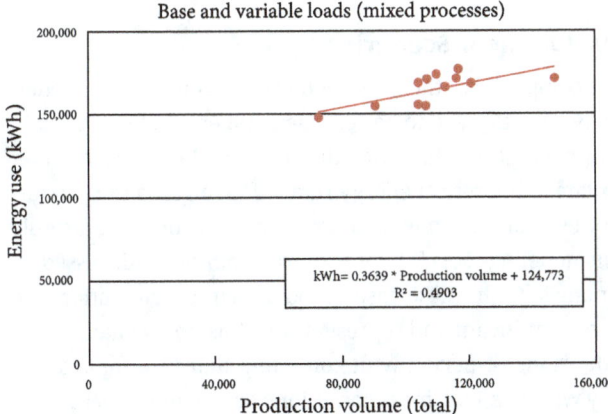

Figure 3.4 PCL for a mixed-process plant combining injection molding and extrusion blow-molding. The diagram reflects a lower correlation ($R^2 = 0.4903$) due to variability in the production mix, emphasizing the need for multifactorial analysis

Figure 3.4 presents the PCL of a plant that combines injection molding and extrusion blow-molding processes. This plant falls into the category of mixed processes, which usually have a more variable process load and will generally exhibit a low correlation coefficient (in this case $R^2 = 0.4903$) due to changes in the product mix and energy intensity of the site average process each month. In these cases, the dependence of consumption ($Energy_{consumption}$) on the monthly compliant production ($Production_{compliant}$) of each process can be modeled using the following equation:

$$Energy_{consumption} \ [kWh] = \sum_{1}^{n} a_i \left[\frac{kWh}{kg} \right] \cdot Production_{compliant_i} \ [kg] + b \ [kWh] \quad (3.4)$$

where:
i is the driving factor
n is the number of driving factors in the EAC
a_i and b have a similar meaning to those of the PCL:
a_i is the variable load for each driving factor i in the EAC
b is the base load for the EAC.

This more generalized multifactorial model is much more difficult to visualize in conventional 2-D plots but can be represented in tabular format. An M&T spreadsheet or other software should handle this calculation perfectly well once the appropriate values of the constants have been found. Multiple regression analysis is one method of establishing these constants and is a function found in spreadsheet programs and dedicated M&T software.

Multiple-factor analysis cannot be easily expressed graphically, but variables can be transformed so that they can be handled using the tools employed for a single factor.

3.2.1.3 Complex or Highly Nonlinear Scenarios

Some industrial processes are so complex that they cannot be represented by the models described above. Different data types may need to be recorded and entered into a mathematical model (specific operating temperatures, batch duration, load weights, number of starts and stops, etc.) to calculate a theoretical energy requirement given the production activity pattern. This is another manifestation of an expected consumption formula (ECF), which can use non-linear modeling tools [3] but is more commonly addressed by developing custom statistical models [9]. In these cases, models require validation and fit data over a period that captures significant and representative business behavior. The literature recommends using the "longest period with consistent data" [3] that can be obtained most quickly and fairly weighs all business conditions or modes of operation.

Although multivariate analysis, statistical methods, or nonlinear modeling can be used to study these complex processes, subdividing the operation into smaller EAC process areas with independent measurement is easier, more straightforward, and more accurate [2]. The most applied practice is to create separate EACs for each driving factor of interest and use linear analysis tools such as PCL to monitor and track the driving factor in question.

3.2.2 Activity-Based Target (ABT)

The PCL represents the average behavior of consumption data vs. compliant production. This implies that there will be data above and below the PCL. The points above the average are the points where energy performance was lower, and the points below indicate better energy resource use. A new regression analysis with these points of better use provides a target, which would be the upgrade line if actions are taken to promote better energy performance than that described by the PCL. This new line is called an activity-based target (ABT).

The analysis using the ABT technique seeks to move future consumption points to a level better than that described by the PCL, which is usually possible without installing any auxiliary energy-saving equipment or implementing significant technology changes. The basis of the method states that it is possible to achieve this since the points that give rise to the ABT have already been obtained during the regular operation of the EAC. Therefore, the aim is to be able to reproduce them consistently.

3.2.3 Performance vs. Efficiency: Specific Energy Consumption Performance Characteristic Curve (PCC)

Equation 3.1 defines the specific energy consumption, which is referred to in this book as net specific energy consumption (SEC_n), since other specific energy consumptions with different meanings and implications will be defined later. From this equation it is possible to calculate the point value of the SEC_n of the EAC. However, from Equation 3.2, which describes the behavior of the PCL, it is possible to intuit that the SEC_n in the EAC also responds to a characteristic curve. Dividing Equation 3.2 by the conforming production, we obtain:

$$SEC_n \left[\frac{kWh}{kg}\right] = a \left[\frac{kWh}{kg}\right] + \frac{b[kWh]}{Production_{compliant}[kg]} \tag{3.5}$$

The analysis of Equation 3.5 allows us to characterize the behavior of the EAC, and four elements stand out:

- SEC_n depends on the compliant production of the EAC, and decreases as the compliant production increases

- The SEC_n behavior is characteristic of the PCL behavior, and therefore if the PCL changes, the SEC_n behavior changes. This implies that the PCL for energy consumption gives rise to a performance characteristic curve (PCC) for specific energy consumption. It is also possible to establish a PCC for the expected behavior of SEC_n from the ABT curve, as shown in Figure 3.5

- The PCC is an asymptotic curve at the value of a, so this value represents the maximum efficiency that it is possible to achieve without intervening in the EAC to improve energy performance

- The variation of SEC_n with conforming production decreases as the base load (b) decreases. With no base load, the SEC_n value is constant.

According to Kent [2], many companies opt for a simple approach to energy management, and calculate the value of net specific energy consumption for each seasonal period as the energy performance indicator (EnPI) or criteria for analyzing EAC information. This approach is erroneous because SEC_n without the connection to production is a misleading indicator and often leads to the wrong conclusions and decisions. As can be seen from Equation 3.1, if there is no significant change in the percentage of non-compliant production, then whenever a lower SEC_n is obtained, a better energy efficiency of the EAC is achieved. However, this does not imply that the EAC has better energy performance. The difference between energy efficiency and energy performance is often confusing. Energy efficiency depends on the point value of SEC_n but energy performance depends on the PCC.

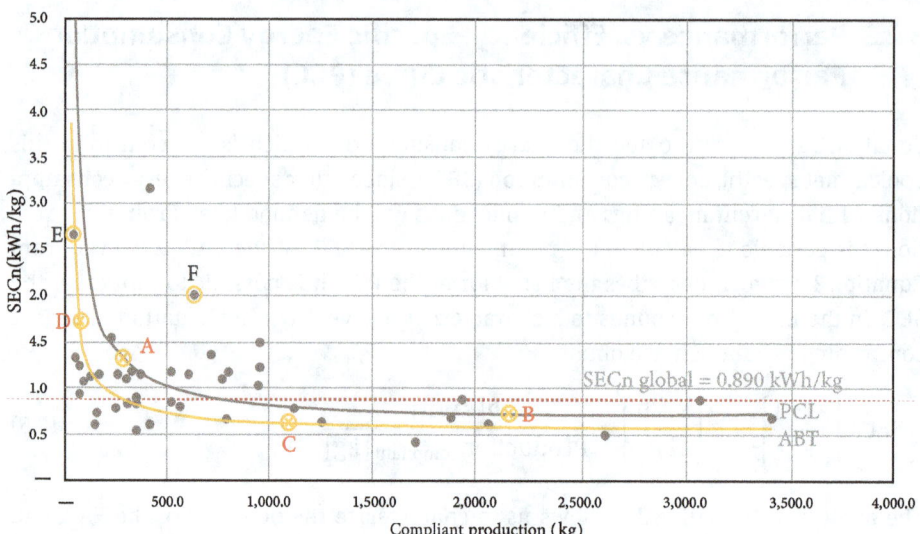

Figure 3.5 Illustration of a PCC derived from both the PCL and ABT in a plastics production process. The PCC reflects how specific energy consumption varies with compliant production, providing insights into operational efficiency and potential improvement thresholds

In comparative terms, a system is more energy efficient the lower the SEC_n. A system exhibits better energy performance if it has a better SEC_n at the same level of compliant product production.

The PCC is a hyperbolic character curve for which the specific energy consumption of the EAC grows very rapidly from a given value of compliant production. This allows analyses that are difficult to obtain from the PCL or ABT. Depending on the seasonal period or the criteria selected for analysis and the EAC chosen, it can provide information to keep SEC_n sufficiently low and thus achieve EAC improvement. These include:

- The minimum size of batches to be assembled by mold, reference, or production order when the analysis criterion is the consolidation of information by mold, reference, or production order, respectively. An example of this application is presented in Section 4.2.12

- The compliant production targets for the EAC, daily, weekly, or monthly, when the analysis of energy and production information is performed in the same seasonal period

- Benchmarking the energy performance of the EAC against the behavior of the industrial sector or subsectors with similar processes

- Validation of the EAC's technological updating decision-making. An example of this is presented in Sections 4.1.1 and 4.1.2.

The equation modeling the PCC of the companies in the sector is:

$$SEC_{n-sector}\left[\frac{kWh}{kg}\right] = 0.649\left[\frac{kWh}{kg}\right] + \frac{10.217\ [kWh]}{Production_{compliant}\ [kg]} \tag{3.6}$$

$R^2 = 0.835$

The equation that models the PCC of the company being benchmarked against the sector is:

$$SEC_{n-company}\left[\frac{kWh}{kg}\right] = 0.643\left[\frac{kWh}{kg}\right] + \frac{30.125\ [kWh]}{Production_{compliant}\ [kg]} \tag{3.7}$$

$R^2 = 0.885$

It is clear from Figure 3.6 that the company's energy performance is lower than the performance of the sector. In both cases, the variable consumption is very similar since the PCC is asymptotic at practically the same value of $a = 0.64$ kWh/kg, so there are no significant differences in performance at high compliant production volumes (approximately above 350 t/month). The difference is in the base load (b), which for the company is 30.125 kWh, while in the sector, it is 10.217 kWh. This means that at low production volumes, the difference is significant. The problem is that the company historically produces less than 350 t/month. Therefore, it should focus on reducing the base load.

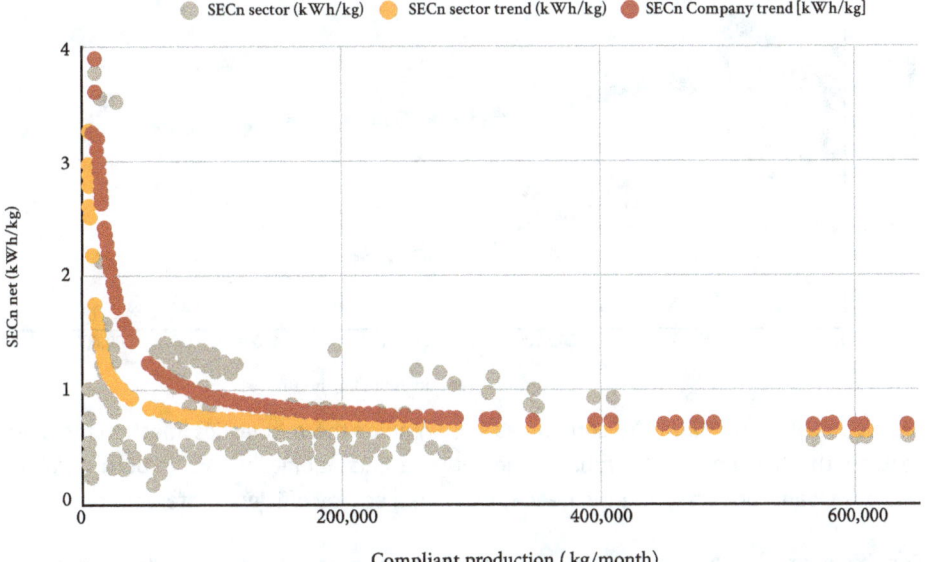

Figure 3.6 Benchmarking of specific energy consumption in the thermoplastics extrusion sector in Colombia, using the PCC from 15 companies. The comparison highlights base load differences and energy performance variability at different production scales

Another example is a similar benchmarking exercise with 15 companies in Colombia's thermoplastic injection molding sector. In this case, the results of this exercise are presented in Figure 3.7. In this case, the equation that models the PCC of the companies in the sector is:

$$SEC_{n-sector} \left[\frac{kWh}{kg} \right] = 1.22 \left[\frac{kWh}{kg} \right] + \frac{35.845 \, [kWh]}{Production_{compliant} \, [kg]} \tag{3.8}$$

$$R^2 = 0.7$$

The equation that models the PCC of the company being benchmarked against the sector is:

$$SEC_{n-company} \left[\frac{kWh}{kg} \right] = 2.0 \left[\frac{kWh}{kg} \right] + \frac{85.245 \, [kWh]}{Production_{compliant} \, [kg]} \tag{3.9}$$

$$R^2 = 0.872$$

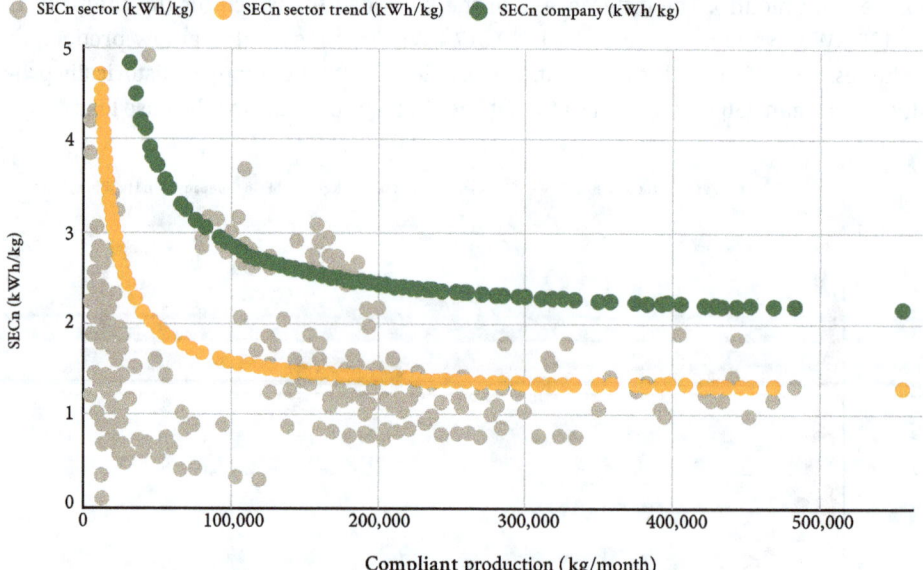

Figure 3.7 PCC comparing 15 companies in Colombia's thermoplastics injection molding sector with a benchmarked company. The figure reveals significant gaps in both variable and base loads, underscoring the need for targeted energy efficiency interventions

For this example, the companies in the sector have a base load of 35,845 kWh and a variable load of 1.22 kWh/kg. The benchmarked company has a higher base load of 85,245 kWh and a variable load of 2.0 kWh/kg. These values indicate that the plant is significantly less efficient and has a lower energy performance than the average company in the sector. The energy intervention required for improvement is at all levels. The company's compliant production is in the range of 80–130 t/month. According to Figure 3.7, the performance curve is highly inefficient for these production levels. To improve energy efficiency significantly, either monthly production must be increased to values above 200 t/month, or the base load must be reduced considerably so that efficiency does not change so rapidly at current production levels. On the other hand, lowering the curve implies improving the plant's energy performance, but this is a more complex procedure requiring a more detailed engineering study.

3.2.4 Cumulative Sum Differences Diagram (CUSUM)

The recursive cumulative sum of residuals (CUSUM) algorithm is one of the earliest methods suggested for statistically detecting changes in engineering processes. It was first introduced as part of steam plant efficiency monitoring during World War II [10]. CUSUM plots represent integrated residuals (rather than raw variance), which is appropriate since energy waste costs are proportional to the total sum, not the variance. CUSUM was first proposed for M&T in UK technical reports (according to [3]) and then described in guidance documents such as that produced by Gotel & Hale in 1989 [11].

The CUSUM chart is a simple but powerful technique for analyzing energy data. It helps with three aspects of energy management:

- Setting challenging targets that are nevertheless demonstrably achievable

- Discriminating sustained periods of adverse behavior, allowing their analysis

- Showing graphically the savings that have been achieved.

The CUSUM value is calculated as the cumulative sum of the differences between actual and expected energy consumption over a given period, typically month to month (although the frequency may vary). This is useful for tracking energy policies and management and determining whether company energy performance changes occur. Using an analogy proposed by Kent [2], CUSUM is similar to a bank balance: it does not matter how much is deposited or withdrawn each month; what matters is what is left in the account at the end of the month.

A CUSUM chart quickly identifies trends and changes in actual versus expected performance. What is essential with CUSUM charts is not the absolute value of the CUSUM; it is the slope of the curve and also any changes in this slope:

- An upward-sloping curve shows that the actual value is consistently higher than the target, i.e. persistent over-consumption

- A falling curve shows that the actual value is consistently below the target, i.e. under-consumption

- A flat (or substantially flat) curve shows that performance is stable and broadly in line with expected values, i.e. operating as expected

- Changes in slope (inflection points) indicate when a persistent change in behavior may have occurred.

The CUSUM chart divides the consumption history into periods of increasing, decreasing, or indifferent energy performance. The causes of each behavior can be discovered by investigating what happened during those periods. In the following subsection, some cases demonstrating the application of CUSUM in the plastics sector are presented.

3.2.4.1 CUSUM Case Analysis

Example A: CUSUM variation from year to year: Effect of adding new machinery

There is the case of a company in Colombia whose main activity is the production of plastic profiles from 100% recycled material and which, at the time of the analysis, has 80 employees. Its metering system is global, so the energy consumption delivered includes all the plant's equipment plus office and general services. The company purchased energy in 2021 and 2022 at an average of $620/kWh. All economic calculations were made using this value.

The goal is to understand how consumption has changed in 2022 compared to the performance of EAC in 2021. To do this, it is necessary to determine the EFC for 2021 consumption. This corresponds to the PCL of EAC in 2021, which has the form of Equation 3.2. Evaluating the PCL with compliant production of 2022 allows for the establishment of estimated consumption based on the performance of EAC in 2021. This value is then compared with actual consumption month by month. The comparison of the estimated consumption with the actual consumption in 2022 is shown in Figure 3.8. The dashed line shows the results of the estimated consumption, while the solid line shows the exact values. If the actual line is below the estimated line, there is better energy performance. This behavior can be seen until May, when the result reverses.

If we take the difference between actual consumption and the estimated consumption month by month, we obtain Figure 3.9. According to the information provided by the company, in May, an agglomerator started operating, which may be responsible for the increase in consumption. Figure 3.10 shows the accumulated sum of the differences (CUSUM), multiplied by the average cost per kWh in 2022, which was $684 COP per kWh. As previously mentioned, an increasing trend over time represents worse performance, a decreasing trend curve shows performance improvement, and a flat trend curve represents stability.

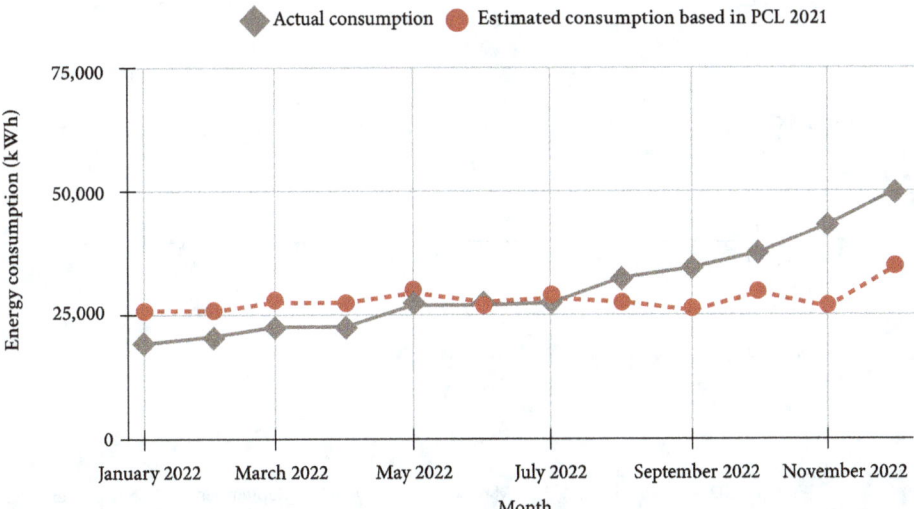

Figure 3.8 Comparison of actual vs. estimated monthly energy consumption in 2022 for a Colombian plastics company. The estimated values are derived using the EAC's 2021 PCL, assessed for compliant production levels for 2022, enabling performance deviation analysis

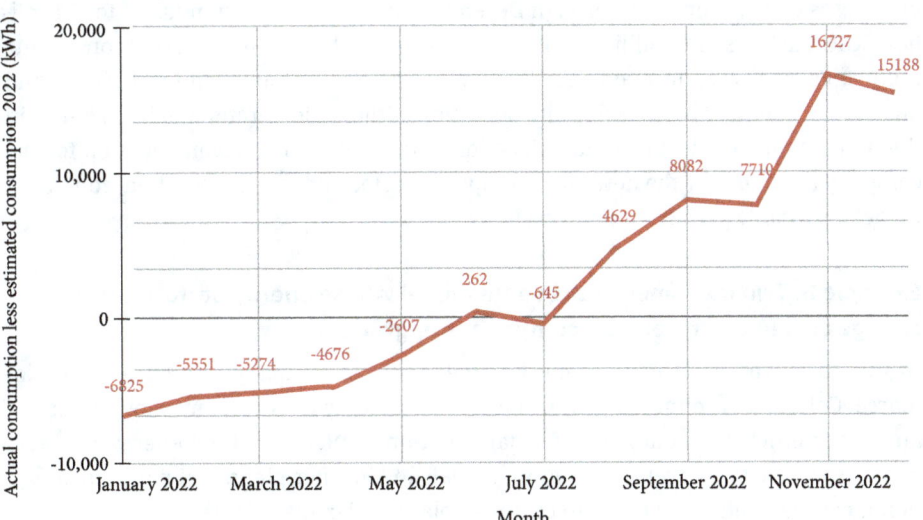

Figure 3.9 Monthly deviations between actual and estimated energy consumption in 2022. The figure highlights changes in performance potentially associated with equipment upgrades and production scheme alteration

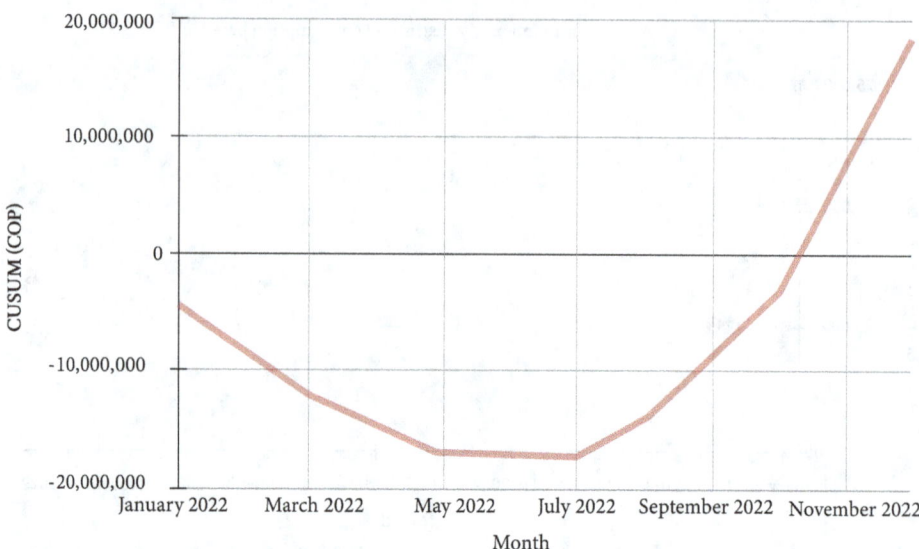

Figure 3.10 CUSUM energy performance analysis in 2022, expressed in Colombian pesos (COP). The figure illustrates accumulated gains and losses compared to the expected consumption based on 2021 data, identifying trends associated with operational changes

Up to May, the company achieved energy savings of approximately 21 million Colombian pesos (COP). From May through December, energy losses accumulated for 31 million COP. A net loss of 10 million COP was generated at the end of the period concerning the plant's energy performance for 2021, with the same production level. This could have been generated by the entry into operation of the agglomerator and by a change in the production scheme due to the relocation of the plant. The recommendation for the company is to monitor the new plant completely relocated and in operation, compared to the most stable period of operation (the year 2021).

Example B: Two machines of the same model whose energy performance changes when one receives a component upgrade

In this case, we performed an energy analysis for a company located in Cundinamarca, Colombia. Preliminary information provided by the company shows a lack of an energy efficiency policy or a formal program implemented for energy management. However, there were consumption goals in kWh per plant, and daily consumption measurements were taken to establish plans and correct deviations.

The company had measuring instruments for each production line and component, where the main components were monitored:

- Injection molding machine
- Chiller

- Dryer

- Compressed air

- Dry cooler

- Others (lightning and services).

Two energy indicators were being tracked, with the data collected in the form of kWh per tonne processed, and USD per tonne processed.

The company was acquiring assets to optimize the plant's energy consumption. The company purchased energy in 2018 at an average of $324/kWh. All economic calculations were performed using this value.

The energy performance of the company during 2018 can be analyzed from the PCL diagram, presented in Figure 3.11, which indicates that:

- The company has a base load of 168,968 kWh, only representing 20.6% of the average 2018 consumption. This is considered an acceptable performance since, for injection companies, it is expected to vary between 20% and 40%. This represents adequate management of the company in terms of the operation of general services, the sizing of the capacities of the plant's equipment, and the use of proper energy practices

- The process load is 0.575 kWh/kg, which is very low (for comparison, a process load less than 1.6 kWh/kg is considered acceptable). It demonstrates that good process parameterization is performed, and highly energy-efficient equipment is available

- The correlation coefficient (R^2) is acceptable, with a value of 0.878.

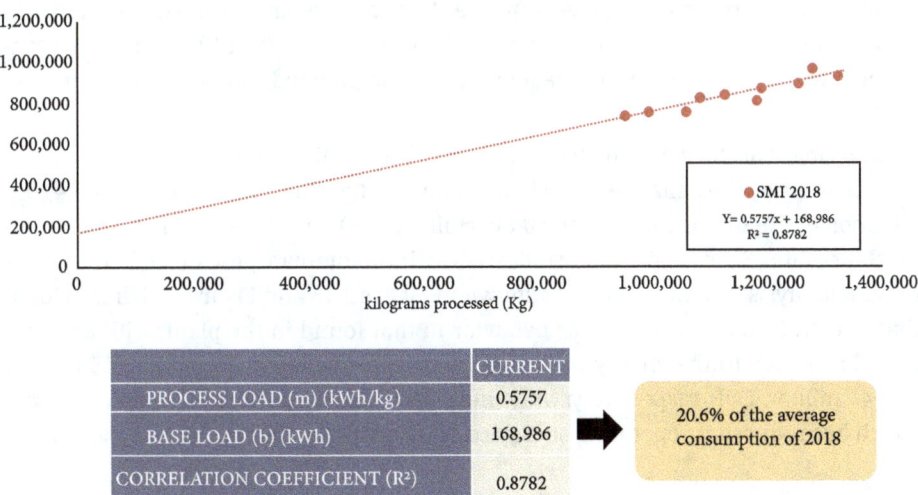

Figure 3.11 PCL diagram of a plant in 2018, showing base and variable energy loads with a high correlation coefficient (R^2 = 0.878). The figure reflects strong performance and process consistency across seasonal production periods

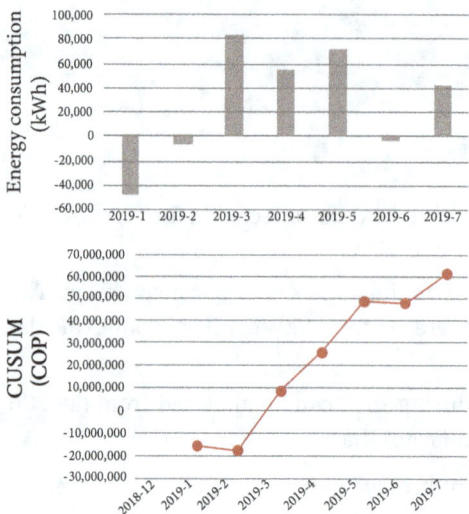

Figure 3.12

(Top) the monthly differences between expected and actual energy consumption; (Bottom) the CUSUM chart for 2019. The figure identifies periods of performance decline and energy losses compared to 2018 targets

The above points to an adequate energy performance of the company and a strong organizational culture that allows the plant to repeat production practices consistently. This facilitates implementing improvements, and allows the sustainability of these improvements to be internalized by the personnel quickly and deeply.

Complementing this first analysis, the differences between expected and actual energy consumption are calculated, and the CUSUM diagram presented in Figure 3.12 (bottom) is used. The figure compares the company's energy performance with the target obtained from the 2018 PCL diagram from January to July 2019. Figure 3.12 shows that the company has a lower energy performance than the previous year and accumulates losses of about 60 million COP (equivalent to approximately 185 MWh of additional consumption). The trend of the 2019 data in the PCL diagram indicates an increase in base load.

Considering that the company has several production lines, the CUSUM analysis can be used to study the behavior of each line individually. The analysis excluded the contribution of the services (compressed air, chillers, etc.) and focused only on the actual machine data. Economic costs are described in Colombian pesos (COP). When the CUSUM analysis is done by lines, as shown in Figure 3.13 and Figure 3.14, it is evident that lines 1, 3, and 4 have similar behavior to that found in the plant, with accumulated losses due to the energy performance of approximately 26 million, 25 million and 42 million COP, respectively. Only line 5 has a better performance than in 2018, which has reduced its operating energy costs by about 28 million COP.

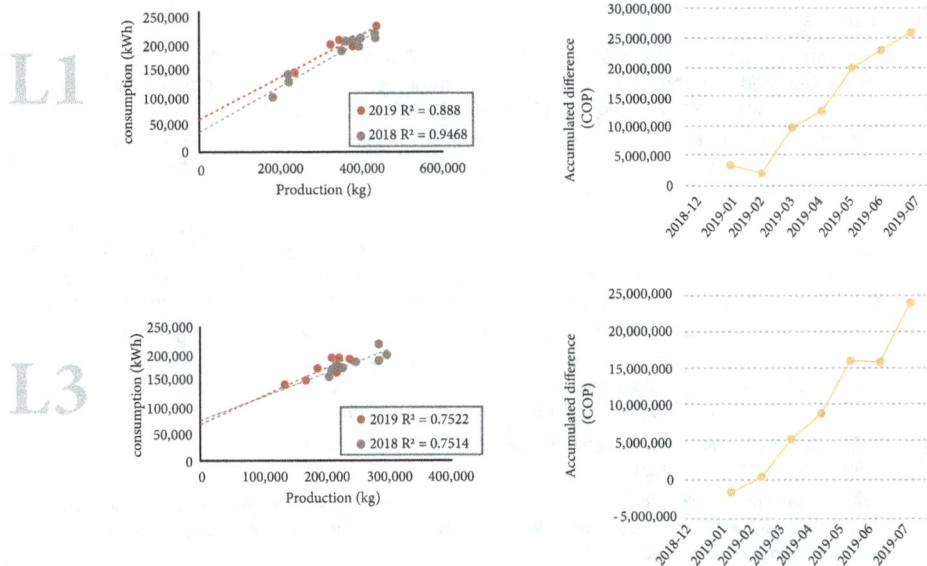

Figure 3.13 PCL (left) and CUSUM (right) diagrams for production lines L1 and L3. The analysis reveals accumulated energy losses in both lines relative to the 2018 baseline

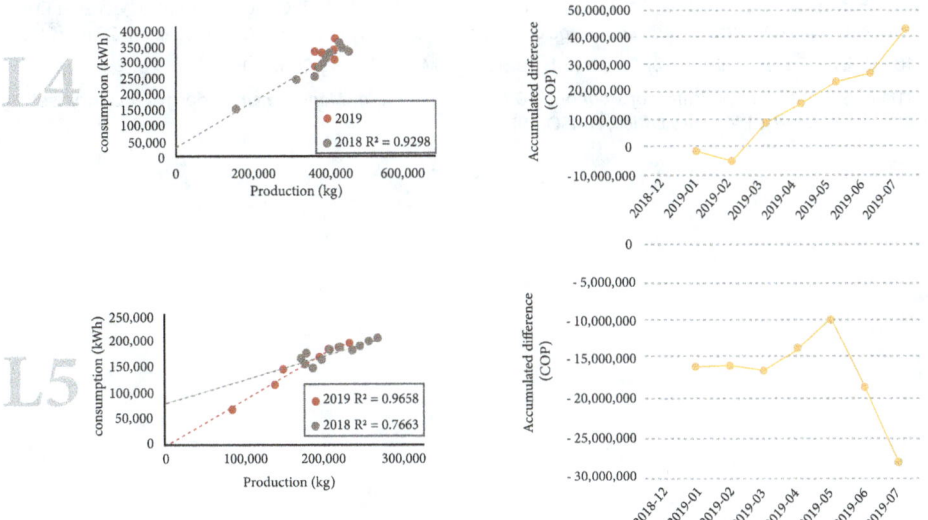

Figure 3.14 PCL (left) and CUSUM (right) diagrams for production lines L4 and L5. While L4 experienced a performance decline, line L5 showed improved efficiency and cost savings compared to the previous year

References

[1] O. A. Estrada-Ramírez, N. A. Muñoz-Realpe, J. A. Patiño-Murillo, F. Chejne, "A novel set of analysis tools integrated with the energy gap method for energy accounting center diagnosis in polymer production", *Resources*, 2025, vol. 14, p. 60, DOI: 10.3390/resources14040060.

[2] R. J. Kent, *Energy Management in Plastics Processing: Strategies, Targets, Techniques, and Tools* [4th edition], British Plastics Federation, 2024, *https://www.bpf.co.uk/Publications/energy-management-in-plastics-processing.aspx*

[3] A. Hilliard, "Energy monitoring and targeting as diagnosis: Applying work analysis to adapt a statistical change detection strategy using representation aiding", Ph.D. Thesis, University of Toronto, 2015, *https://www.researchgate.net/publication/292608832_Energy_Monitoring_and_Targeting_as_diagnosis_Applying_work_analysis_to_adapt_a_statistical_change_detection_strategy_using_representation_aiding*

[4] O. Estrada, J. C. Ortiz, A. Hernández, I. López, F. Chejne, M. del Pilar Noriega, "Experimental study of energy performance of grooved feed and grooved plasticating single screw extrusion processes in terms of SEC, theoretical maximum energy efficiency and relative energy efficiency", *Energy*, 2020, vol. 194, article no. 116879, DOI: 10.1016/j.energy.2019.116879

[5] K. Bhattacharjee, *Industrial Energy Management Strategies: Creating a Culture of Continuous Improvement*, River Publishers, 2020

[6] O. A. Estrada Ramírez, "Estudio de la influencia del proceso de plastificación en la eficiencia energética del proceso de extrusión monohusillo", Ph.D. Thesis, Universidad Nacional de Colombia, 2021, *https://repositorio.unal.edu.co/handle/unal/79375*

[7] O. Estrada, I. D. López, A. Hernández, J. C. Ortíz, "Energy gap method (EGM) to increase energy efficiency in industrial processes: Successful cases in polymer processing", *Journal of Cleaner Production*, 2018, vol. 176, pp. 7–25, DOI: 10.1016/j.jclepro.2017.12.009

[8] Y. Ban, "Energy decision making of steel company based on energy management system", *IFAC-PapersOnLine*, 2020, vol. 53, pp. 608–613, DOI: 10.1016/j.ifacol.2021.04.151

[9] "Energy savings toolbox: An energy audit manual and tool", CIPEC, Government of Canada Publications, 2008, *https://publications.gc.ca/site/eng/9.856168/publication.html* [accessed 29 July 2024]

[10] O. Lyle, *The Efficient Use of Steam* (6th edition), H.M. Stationery Office, 1947

[11] D. G. Gotel, *The Application of Monitoring & Targeting to Energy Management* ("Energy Efficiency" series no. 8), Energy Efficiency Office, HMSO, 1989

4

The Energy Gap Method (EGM)

Omar Estrada, Iván López

Among the methods and/or methodologies that can be used to support a company's energy management, we highlight two: the methodology of monitoring and targeting by objectives (M&T) and the energy gap method (EGM). The two methods are complementary, since while M&T is an energy consumption control methodology for the continuous improvement of performance and energy efficiency of the energy accounting center (EAC), the EGM is a diagnostic method of the EAC that seeks to determine the causes of such performance and efficiency, in addition to providing a framework under which areas for improvement are more easily identified. Actions are prioritized to achieve the improvement objectives. The differences between M&T and EGM can be better understood from Table 4.1.

The M&T method offers the possibility of understanding and improving the energy performance of an EAC by continuously replicating the best behaviors shown by the EAC and avoiding repeating the actions that will enhance its energy performance.

M&T successfully presents a view of the energy performance of the EAC using very little information and allows comparisons between different measurement periods to evaluate improvements. However, there are questions that the method itself cannot answer. Questions such as:

- What is the inefficiency of the process?

- What is the potential for improvement?

- Where should the inefficiencies of the production process be addressed as a priority?

It is also common for companies to divide the operation by areas or management through various committees that manage their indicators. Production, process, quality, and technology management areas or committees are typical in most companies; they all work most of the time, separate from each other. This raises a new question:

- Could the continuous improvement of an industrial plant be integrated from a methodology with common indicators for all areas?

Table 4.1 Characteristics and Differences Between the M&T and the EGM

M&T	EGM
A control method	A diagnostic method
Analysis is performed for selected EACs	Analysis is performed for selected EACs
Uses the determination of an EAC's performance characteristic line in the past to define a target	Uses monitoring and targeting tools such as the performance characteristic line (PCL), the activity-based target (ABT), and the performance characteristic curve (PCC) implemented in an EAC to determine its current energetic state
The objective is determined based on the ABT, the PCC, and the cumulative sum (CUSUM) diagrams, which permit consumption behavior to be established	Determines energy gaps and detects production elements that require specific actions
Analysis is performed for given periods, where the information on consumption and production is obtained at regular intervals, which can be days, weeks, or months	The analysis is carried out using criteria such as products, molds, references, materials, heads, and machines, so the analysis periods are not regular
The tools are used within regular intervals to perform continuous and comparative consumption analysis, allowing early decisions to be made and ensuring compliance with the goals	The tools are used when it is necessary to determine the causes of a specific EAC performance or observed efficiency, define actions, and ensure compliance with the goals

The EGM [1] can be an option to answer this type of question. It is a method complemented with M&T and benchmarking principles to analyze EAC consumption and define priority intervention areas, as presented below.

Energy management systems are usually built from general to particular. Initially, the EAC is the whole company. As the management system evolves, it is particularized by performing analyses by plants up to the point of performing energy analyses by production lines. M&T is a more than sufficient methodology for more general studies. The EGM has shown its value when the energy management system evolves and analysis by production lines is required. As the energy management system matures, it is more difficult to find improvements in the proper sizing and adjustment of general services, good energy consumption practices, and insulation to avoid energy losses due to interaction with the surroundings, among others. At this point, EGM becomes a critical methodology to support improvement, which requires entering with technical criteria and deep knowledge of the specific production processes of the plant.

4.1 Preliminary Concepts and Method Definitions

Energy inefficiency at the industrial level comes from five essential sources: production, quality, processes, technology, and/or state of the art or technique. Of the five sources, the first four are controlled by the company producing the goods. At the same time, the last one depends on improving state-of-the-art technology through R&D processes, which equipment manufacturers, universities, technology development centers, and research institutes usually carry out.

The EGM has the potential to help identify the sources of energy inefficiency. It defines six levels of specific energy consumption (SEC). Between two SEC levels, an energy gap (EG) is established, giving rise to five defined energy gaps: the production energy gap, the quality energy gap, the process energy gap, the technology energy gap, and the R&D energy gap. In this way, the energy gaps are aligned with the sources of energy inefficiency and, therefore, reflect the inefficiency related to each factor. Figure 4.1 illustrates the EGM energy gaps. Some preliminary concepts and definitions will be introduced before focusing on a detailed study of each.

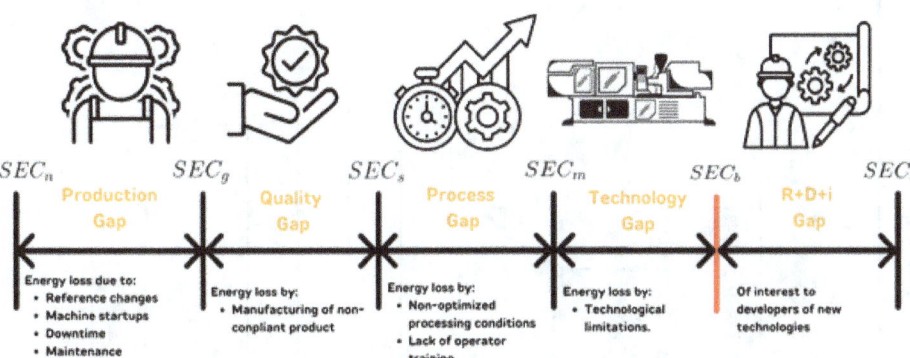

Figure 4.1 EGM applied to polymer processing. This figure illustrates the EGM, which decomposes the specific energy consumption (SEC) into five sequential improvement gaps: Production, Quality, Process, Technology, and Research & Development & Innovation (R&D&I). Each gap represents a category of energy loss – from operational inefficiencies such as equipment startup and downtime through product nonconformity and suboptimal processing conditions to inherent technological limitations. The final gap highlights areas of opportunity for innovation and the development of new technologies to push the boundaries of energy efficiency in polymer production systems

4.1.1 Net Specific Energy Consumption (SEC$_n$)

It is the highest level of specific energy consumption and the one that meets the classical definition of SEC, which is:

$$\text{SEC}_n \left[\frac{\text{kWh}}{\text{kg}}\right] = \frac{\text{Energy}_{\text{consumption}} \, [\text{kWh}]}{\text{Production}_{\text{compliant}} \, [\text{kg}]} = \frac{E_a}{W_c} \tag{4.1}$$

The most important consideration is that energy consumption (Energy$_{\text{consumption}}$) and compliant production (Production$_{\text{compliant}}$) must correspond to the same time period. When the EAC corresponds to the entire plant, consumption can be obtained directly from the energy bill. In another case, energy consumption is obtained by performing an integration by Simpson's method, as shown in Figure 4.2 and Equation 4.2.

$$\text{Energy}_{\text{consumption}} \, [\text{kWh}] = \frac{\sum_{i=1}^{n} \left(\text{Power}_{\text{demand}_i} + \text{Power}_{\text{demand}_{i-1}}\right)}{2} \cdot (t_i - t_{i-1}) \tag{4.2}$$

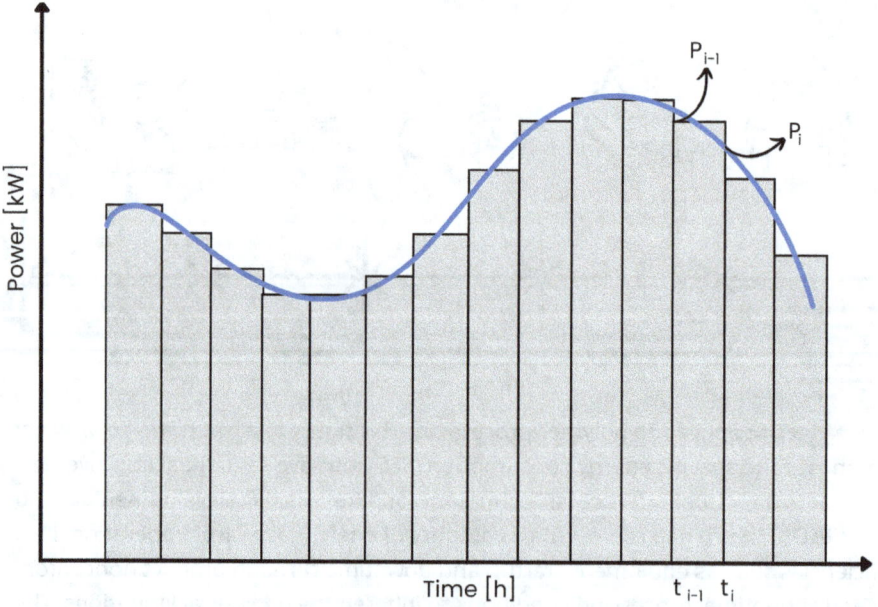

Figure 4.2 Discretization of the area below the curve for determination of energy consumption by Simpson's method of integration of the power demand vs. time curve. The figure illustrates how total energy consumption is calculated as the area under the power–time curve, supporting accurate SEC determinations

It should be remembered that the SEC_n as a single value may be an unreliable indicator for energy analysis of the EAC, since it depends on production. Higher production results in a more energy-efficient process but does not necessarily imply better performance for the EAC, as is evident from the M&T analysis (see Chapter 3).

The behavior of the EAC, in terms of energy consumption, can be obtained by applying M&T, for which it is necessary to consolidate the consumption and production information in equal periods. It must then be decided whether work shifts, days, weeks, or months are used to carry out the analysis. In this way, the PCL, the ABT, and the PCC of the EAC can be obtained. In this case, the base load will be those consumed while the EAC is not producing, either by shift, day, week, or month, according to the chosen period. This way facilitates the control of EAC consumption and the comparative analysis using the CUSUM diagrams to show its performance and energy efficiency improvements. However, this analysis method does not provide information on what is going right or wrong with the EAC. An alternative way of using M&T tools developed for EGM to complement and strengthen the diagnosis will be presented below.

4.1.2 Thermodynamic Specific Energy Consumption (SEC_t)

The SEC_t is the specific energy required to transform 100% of the raw materials into products in an ideal condition of absence of energy losses. The SEC_t can be obtained from a black box energy balance in the EAC, as shown in Chapter 2. This balance considers all the changes of internal, kinetic, and potential energies in the mass entering the black box. As previously presented, kinetic and potential energy changes are negligible in polymer processing, so the specific thermodynamic energy consumption is expressed as reproduced below in Equation 4.3.

$$SEC_t = \frac{\Delta e_u}{W_t} \tag{4.3}$$

Equation 4.3 is the idealized and simplified thermodynamic-specific energy consumption (SEC_t). No system can achieve 100% energy efficiency due to entropy generation. For this reason, when studies are available at the maximum energy efficiency level for a particular industrial process, it is recommended to use that way of determining it. Otherwise, Equation 4.3 is the best option. In recent years, there has been an increased interest in modeling energy efficiency in industrial processes, using energy, exergy, and entropy balances [1] to determine the maximum energy efficiency value that is thermodynamically possible to obtain from a process or technology. Figure 4.3 shows that the difference between SEC_n and SEC_t represents the total energy inefficiency due to energy loss occurring in the EAC for any reason.

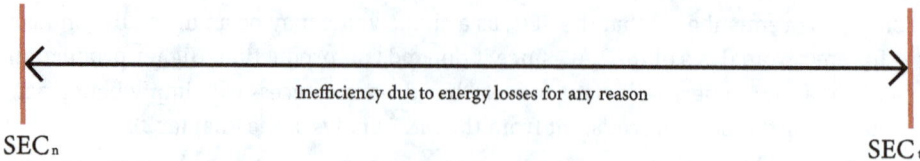

SEC$_n$ SEC$_t$

Figure 4.3 Inefficiency in EAC energy consumption due to total energy losses. The chart visualizes the gap between net and thermodynamic SEC, highlighting the total inefficiencies due to all forms of energy loss within the system

4.1.3 Stable Specific Energy Consumption (SEC$_s$)

The stable specific energy consumption is obtained when the process reaches a stable and continuous operating condition over time. The SEC$_s$ represents the minimum SEC that is possible to get, at the usual operating conditions, if all the mass processed is converted into compliant product and when there are no starts, stops, or idle times during the measurement period. In this case, the specific steady-state energy consumption responds to the following equation:

$$SEC_s \left[\frac{kWh}{kg} \right] = \frac{AVGPower_{demand} \ [kW]}{\dot{m} \left[\frac{kg}{h} \right] \cdot \left(1 - f_{inherent \ scrap} \right)} \tag{4.4}$$

where:

$\dot{m}$ is the mass flow

$f_{inherent \ scrap}$ is the fraction of scrap inherent to the process

Inherent scrap is usually, within specific reasonable ranges, unavoidable. For example, in producing sheet and flat film, the edge trim must be removed, constituting 5–10% of the entire melt to be extruded. When producing containers by extrusion blow molding, the pinch-off areas represent 15–50% of the preform weight. When parts are injected with cold runner technology, the feed channels must be removed and can constitute 3–10% of the mass being injected. In the thermoforming process, 30–50% of the sheet must be reprocessed in the extrusion process, since there is always considerable skeletal waste when the product is die-cut.

On the other hand, AVGPower$_{demand}$ it is the average power demand of the process during a period in which the process is in a steady state or quasi-steady state. To determine it, it is not necessary to perform measurements for long periods if it is ensured that the measurement period is representative of the steady state behavior of the production line.

In continuous processes such as the extrusion process, it is easy to recognize these periods of stable operation through the demand over time, as shown in Figure 4.4. This figure identifies two periods of stability with an average power demand of around 21 kW.

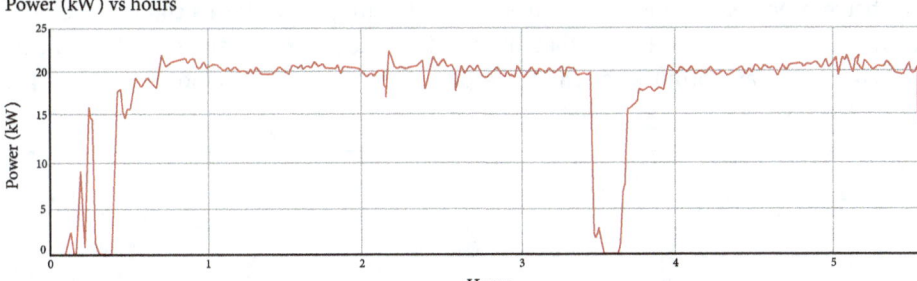

Figure 4.4 Power demand over time in a vulcanized rubber extrusion process. The plot identifies stable operating periods used to estimate steady-state energy consumption in continuous production

In the case of steady-state continuous processes:

$$\dot{m} \left[\frac{kg}{h} \right] = \text{Output} \left[\frac{kg}{h} \right] \tag{4.5}$$

Stability is more challenging to recognize when it comes to discontinuous processes such as molding, which occur in batches. Within the cycle, the demand varies depending on the stage of the cycle. It does not require the same power to open or close the mold as it does to plasticize the polymer, as seen in Figure 4.5.

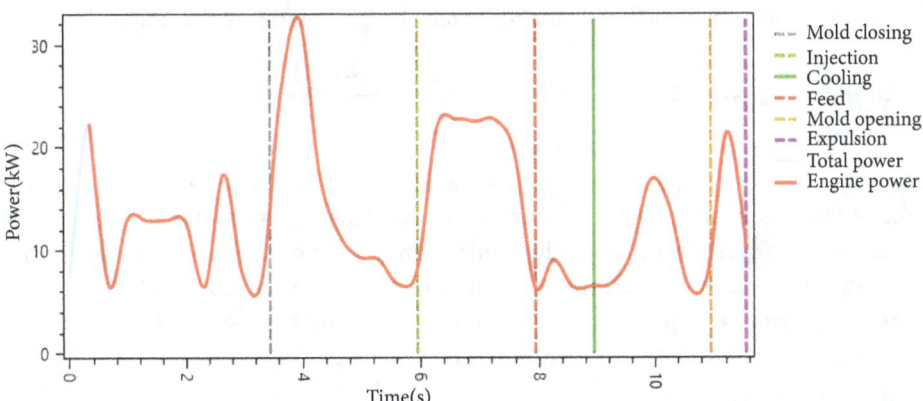

Figure 4.5 Energy demand profile during a typical injection molding cycle. The figure shows variations in power requirements for each stage of the injection cycle, revealing the variable nature of demand for discontinuous processes

The batch processes do not reach a steady state, but they do achieve a condition known as quasi-stationary. This implies that, in this state, the pressure, temperature, velocity, and energy demand profiles have a similar pattern on each cycle, as shown in Figure 4.6.

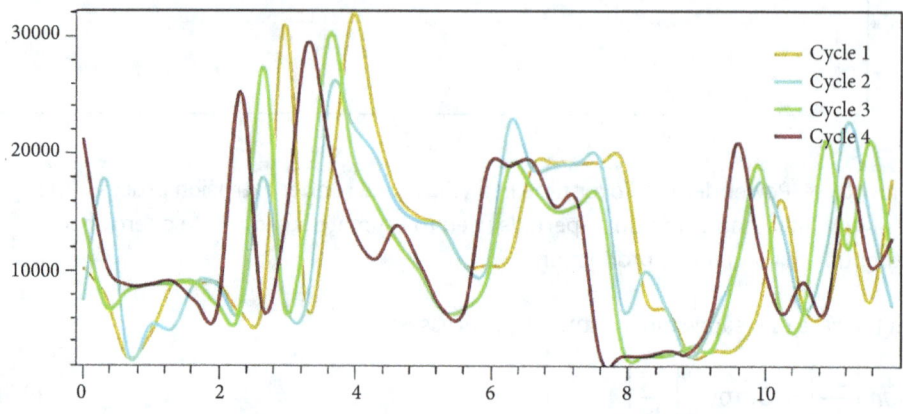

Figure 4.6 Power demand fluctuations in a batch polymer processing operation. The quasi-steady-state behavior observed in repeated cycles allows estimation of average energy consumption per cycle

Because of this demand behavior, the average power demand cannot be calculated as the arithmetic average of the power in the selected stability or quasi-stability time range. In this case, the average power demand between time t_1 and time t_2 is computed as:

$$\text{AVGPower}_{\text{demand}}[\text{kW}] = \frac{\text{Energy}_{\text{consumption}_{t_1 \rightarrow t_2}}[\text{kWh}]}{(t_2 - t_1)[\text{h}]} \tag{4.6}$$

Molding processes in polymer processing are typically fast, with each cycle lasting from a few seconds to usually less than a minute. This implies that when demand is measured at a given periodicity, the demand may vary between one measurement time and the next unless an integer number of cycles are executed within the measurement period. This problem is minimized if the stability time between t_1 and t_2 is large enough to include several cycles or batches.

In discontinuous processes:

$$\dot{m}\left[\frac{\text{kg}}{\text{h}}\right] = \frac{\text{Mass}_{\text{total}_{\text{per cycle}}}[\text{kg}]}{t_{\text{cycle}}[\text{h}]} \tag{4.7}$$

where:

$\text{Mass}_{\text{total}_{\text{per cycle}}}$ is the total mass processed per cycle
t_{cycle} is the cycle time.

The determination of SEC_s allows division of the total energy inefficiency of the EAC into two components: the one due to the base load and the one related to the variable load, as shown in Figure 4.7. All actions taken to reduce the base load bring SEC_n closer to SEC_s, and all actions that improve the energy performance of the technology brings SEC_s closer to SEC_t.

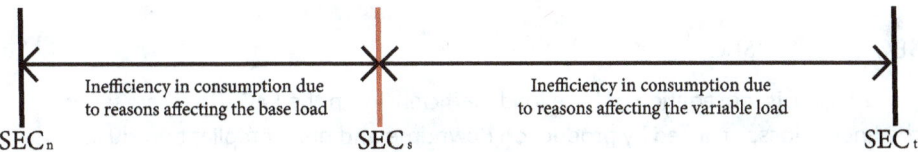

Figure 4.7 Segmentation of energy inefficiency in the EAC into base load and variable load components. This differentiation helps guide efforts to reduce idle consumption and improve operational performance in the efficiency related to fixed or based load and variable load

4.1.4 Gross Specific Energy Consumption (SEC$_g$)

From the definitions presented, the origin of the differences between the SEC_n and SEC_s begins to be intuited. In principle, these two values are not equal because the process does not work stably, producing compliant products during 100% of the operation time. This implies that downtime and the investment of time and energy in the production of non-compliant product increase the SEC_n value. The question is: is it possible to determine how much energy is invested in producing non-compliant products during production downtime? To achieve this, the gross specific energy consumption or SEC_g is calculated, which responds to the following equation:

$$SEC_g \left[\frac{kWh}{kg} \right] = \frac{SEC_s \left[\frac{kWh}{kg} \right]}{(1 - f_{non-compliant\ product})} \tag{4.8}$$

Where $f_{non-compliant\ product}$ is the fraction of non-compliant product produced, taking into account not to include the fraction of scrap inherent to process ($f_{inherent\ scrap}$) or the scrap material generated during the startup, shutdown, or reference change processes. These three make up all the scrap material in the production process, so it is advisable to keep their accounting separate. For this reason, it may be necessary to review the information systems, to have the actual values of the fraction of non-compliant products produced in each production order or during each period selected for the information analysis.

Figure 4.8 schematizes how, with SEC_g, it is possible to divide the inefficiencies that affect the base load in the EAC into the inefficiency associated with non-compliant production and the inefficiency related to lost production time, while the equipment keeps consuming energy.

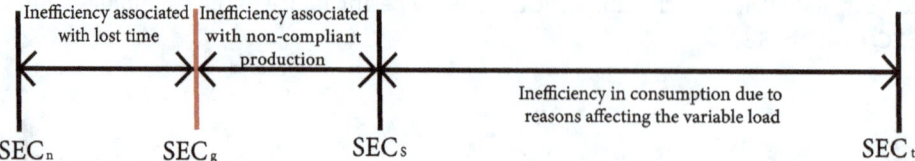

Figure 4.8 Disaggregation of base load inefficiencies in the EAC. The chart separates the energy losses caused by production downtime and non-compliant products, enhancing diagnostic precision for process improvement

4.1.5 Machine Specific Energy Consumption (SEC$_m$)

The machine specific energy consumption or SEC$_m$ is the minimum value of SEC$_s$ or stable specific energy consumption, when the process speed, and hence the mass flow, is maximized, on the specific machine or EAC under study. Thus:

$$SEC_m \left[\frac{kWh}{kg} \right] = SEC_s \left[\frac{kWh}{kg} \right] \text{ at } \dot{m} \to \dot{m}_{max} \tag{4.9}$$

Usually, when measuring the energy consumption of the EAC, the data corresponds to the usual operating conditions of the processing equipment, which in most cases are not the operating conditions that maximize the processing speed, taking care that there is no increase in the non-compliant product. When processing the energy information, it is possible to estimate how much the SEC$_s$ would be reduced when increasing $\dot{m}$ (as will be verified later in Figure 4.24 and Equation 4.30). However, applying process optimization procedures requires knowing how much $\dot{m}$ could be increased. Only after that is possible to determine the actual value of SEC$_m$. The SEC$_m$ is not a unique value, since the maximum process speed is reached at a different condition, depending on multiple factors that rely on the EAC technology, the tooling used and how this technology interacts with the other materials it works with. However, the lowest SEC$_m$ that the technology associated with the cost center can reach is:

$$SEC_m \left[\frac{kWh}{kg} \right] \to a \left[\frac{kWh}{kg} \right] \text{ at } \dot{m} \to \infty \tag{4.10}$$

Based on Equation 4.10, the discussion of the concept of energy efficiency will be expanded in Section 4.1.14. As established in Chapter 3, the variable load is associated with the energy performance of the technology. The SEC$_m$, as shown in Figure 4.9, makes it possible to establish whether the impact of variable load due to the work of the technology when operating far from optimal process conditions and the effect due to the use of inherently inefficient technology.

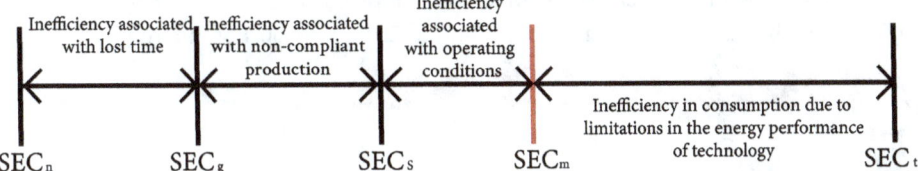

Figure 4.9 Analysis of variable load inefficiencies in EAC energy consumption. This figure classifies energy losses into those caused by suboptimal operating conditions and those due to limitations in technological efficiency

4.1.6 Benchmark Specific Energy Consumption (SEC$_b$)

The SEC$_b$ is the SEC$_m$ value of a technology selected as a reference and with which the SEC under study is to be compared. According to Peterson & Belt [2] there are three types of benchmarking: industrial benchmarking, historical benchmarking, and internal company benchmarking. For the determination of SEC$_b$, only industrial and internal company benchmarking are applicable. In other words, SEC$_b$ is the SEC$_m$ of a technology against which we want to compare our process technology. When comparing the company's technology with the state of the art, the SEC$_b$ is the SEC$_m$ of the best available technology. This is known as industrial benchmarking and is the most widely used. When comparing the energy performance of a process against the best process within the company, the SEC$_m$ of the company's best technology is used as a reference, and this is known as internal benchmarking.

$$SEC_b \left[\frac{kWh}{kg} \right] = SEC_m \left[\frac{kWh}{kg} \right] \text{ (from reference technology)} \qquad (4.11)$$

For industrial benchmarking, the machine-specific energy consumption value of the best available state-of-the-art technology is used as a reference. This requires constant technological surveillance, visiting trade fairs, and permanent monitoring of the technological advances of the leading machinery manufacturing companies. Some sources of this information can be found in scientific articles, norms or standards, and even catalogs of technologies like the EAC under study. For example, the European Association for Plastics and Rubber Machinery Manufacturers (EUROMAP) defined a classification of ten SEC levels for injection molding machines (EUROMAP 60.1 [3]) and for extrusion blow-molding machines (EUROMAP 46.1 [4]), with the intention that they serve as energy efficiency labels. This entity represents about 500 companies that manufacture machinery for the plastics and rubber industry. EUROMAP 60.1 establishes the ten SEC levels for injectors with plasticizing units larger than 25 mm, as shown in Table 4.2.

Table 4.2 Values for Class Classification of Thermoplastic Injection Molding Machines, in Terms of Their Energy Efficiency, According to EUROMAP 60.1 Standard [3]

Class	SEC_m [kWh/kg]
1	>1.50
2	≤1.50
3	≤1.20
4	≤0.96
5	≤0.77
6	≤0.61
7	≤0.49
8	≤0.39
9	≤0.31
10	≤0.25

To use these values in Table 4.2, the measurement procedure established in the standard must be taken into account, since there will be differences as a result of the existing differences between the usual operating conditions of the injection molding machine under study and the conditions in which the standard proposes to measure energy efficiency in terms of SEC. For example, EUROMAP 60.1 [3] suggests that the measurement be carried out with virgin polypropylene with a melt flow rate (MFR) of 20–25 g/10 min (230 °C/2.16 kg) without drying or preheating at ambient temperature (below 30 °C). More approximate values can be obtained by correcting the values after operating the injection molding machine to the conditions established in the standard with reference material and customarily used material. In the absence of better reference values, it is recommended to use those in Table 4.2.

Concerning hollow body extrusion-blowing equipment, EUROMAP 46.1 [4] establishes reference-specific energy consumption levels for energy labeling, presented in Table 4.3, which are independent of the diameter of the plasticizing unit. In this case, it is also recommended to validate the measurement procedure established in the standard, which, among others, establishes that the documented values serve as a reference when processing high-density polyethylene with a density of 0.950 ± 0.005 g/cm^3 and an MFR of 8–10 g/10 min (190 °C/21.6 kg).

Table 4.3 Values for Class Classification of Extrusion Blow-Molding Machine, in Terms of Energy Efficiency, According to EUROMAP 46.1 Classification [4]

Class	SEC_m [kWh/kg]
1	>1.30
2	≤1.30
3	≤1.00
4	≤0.80
5	≤0.62
6	≤0.53
7	≤0.45
8	≤0.39
9	≤0.34
10	≤0.29

Unfortunately, no such standardization or comparison initiatives for other types of polymer processing technologies have been found at the time of writing. Another source for industry benchmarks is technology trade shows, where leading companies in developing polymer processing equipment exhibit their latest technological advances and make claims of productivity and energy efficiency that can be used. These reference values constantly change as technology evolves and is adopted, so continuous technological monitoring of this issue is essential.

Regardless of the source, for the determination of SEC_b, the fraction of scrap inherent in the process ($f_{inherent\ scrap}$) must be considered. For example, the corresponding EUROMAP standards referenced previously do not consider the burrs generated by the pinch-off in the extrusion-blowing process, nor the scrap generated by the distribution channels in injection molding with cold runner molds.

Industrial benchmarking is used when the company wants to study a potential investment in a new technology or compare the energy performance of its technology with the best technology in the state of the art. In contrast, the global benchmarking of the company is used when it wants to evaluate moving the manufacturing process to the reference production line, investing in technology like a reference one, or assessing the differences in the energy performance of similar technologies to improve it. These comparison values don't consider that technologies may be using general-purpose services, and depending on how the measurement is done, this factor can distort the comparison. However, it is still valuable for benchmarking purposes, comparing the performance of different technologies against a single benchmark.

The scheme presented in Figure 4.10 shows that inefficiencies in consumption due to limitations in the energy performance of the technology can be divided into two large groups, separated by SEC_b. The first group contains the inefficiencies derived from using technologies with lower performance than the best existing technologies in the state of the art. The second group is the inefficiencies associated with the non-existence of better-performing technologies in the state of the art.

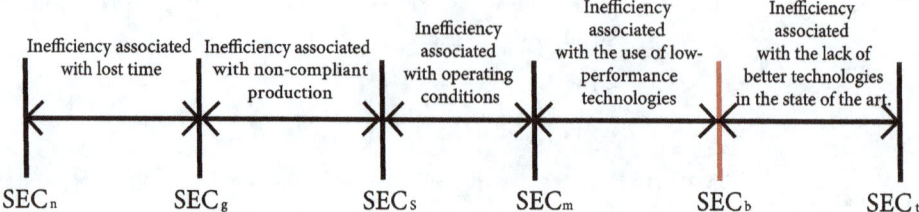

Figure 4.10 Classification of energy inefficiencies in the EAC due to technological limitations. The figure divides the total inefficiency into segments attributed to outdated technologies and the absence of higher-performance alternatives, highlighting areas of opportunity for technological renewal and investment prioritization

4.1.7 Hierarchy of Specific Energy Consumption Levels

In the EGM, there is a hierarchy of the different definitions of specific energy consumption, as shown in Figure 4.10. From this hierarchy appears the first rule to be fulfilled in the EGM:

$$SEC_n \geq SEC_g \geq SEC_s \geq SEC_m \geq SEC_b \geq SEC_t \qquad (4.12)$$

As previously stated, the SEC_n – SEC_t difference is the energy gap between the actual EAC performance and the ideal performance, when the energy efficiency is 100%. This difference represents the energy inefficiency of the EAC, so it is associated with the magnitude of the energy lost for different reasons, as shown in Figure 4.3.

The concept of energy gap was introduced by Hirst & Brown [5] in 1990 and presented by Schulze et al. [6] in 2016 in their review of previous findings and the conceptual framework of energy management in industry. In this work, the energy gap is defined as the difference between the actual and theoretical levels of energy efficiency that can be achieved when cost-effective technologies are implemented. EGM uses the differences between specific energy consumption values between two consecutive levels to define energy gaps [1], which are presented below.

4.1.8 Production Energy Gap (EG$_{production}$)

The production energy gap represents the increase in specific energy consumption generated by energy losses and, therefore, the inefficiencies associated with unscheduled downtime, reference changes, and the energy expenditure required to transform materials during startup and shutdown processes and reference changes while the process stabilizes. This gap is calculated with Equation 4.13 and is presented in Figure 4.11.

$$EG_{production} = SEC_n - SEC_g \qquad (4.13)$$

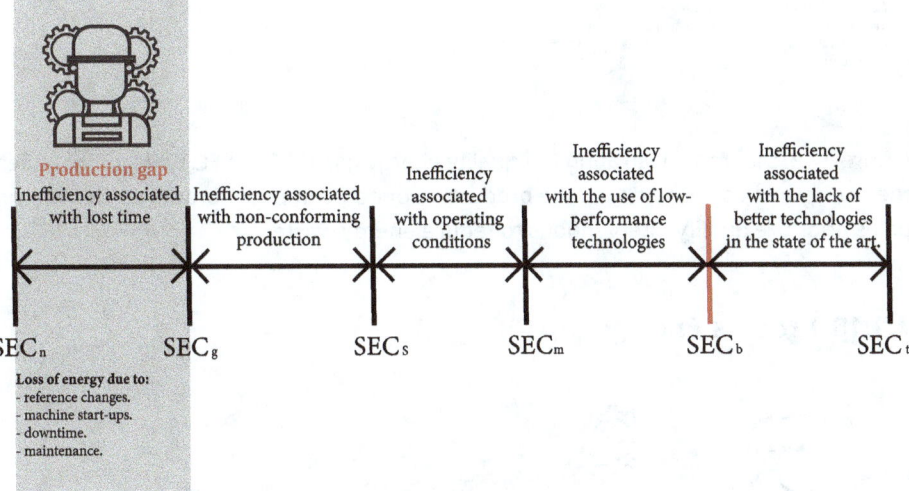

Figure 4.11 Visualization of the production energy gap, defined as the difference between the net and gross specific energy consumption (SEC$_n$ – SEC$_g$). This gap reflects energy loss due to downtime, startup, shutdowns, and changeovers, highlighting inefficiencies in production continuity

4.1.9 Quality Energy Gap (EG$_{quality}$)

The quality energy gap represents the increase in specific energy consumption and the inefficiencies generated by the energy losses associated with the energy spent producing non-compliant products. This gap is calculated with Equation 4.14, represented in Figure 4.12.

$$EG_{quality} = SEC_g - SEC_s \qquad (4.14)$$

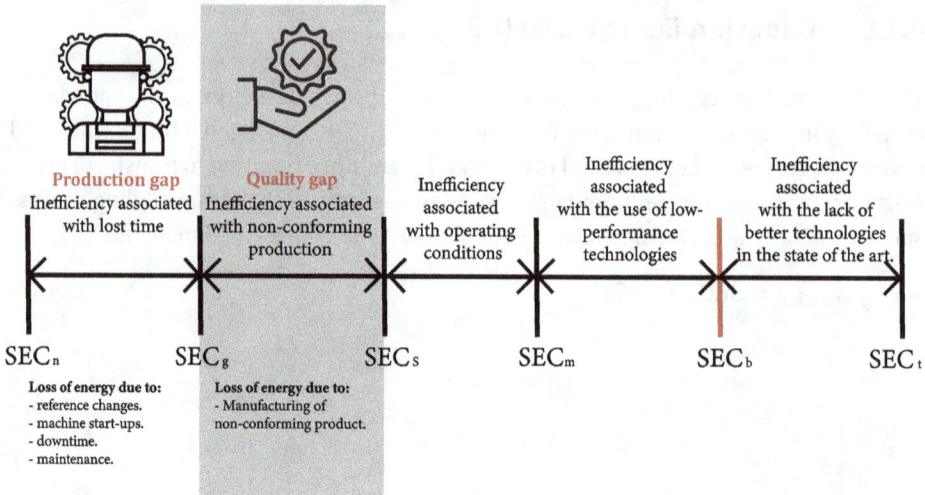

Figure 4.12 Diagram illustrating the quality energy gap (SEC_g – SEC_s), which quantifies the energy loss associated with the production of non-compliant products. This metric helps prioritize quality interventions to reduce energy waste

4.1.10 Process Energy Gap ($EG_{process}$)

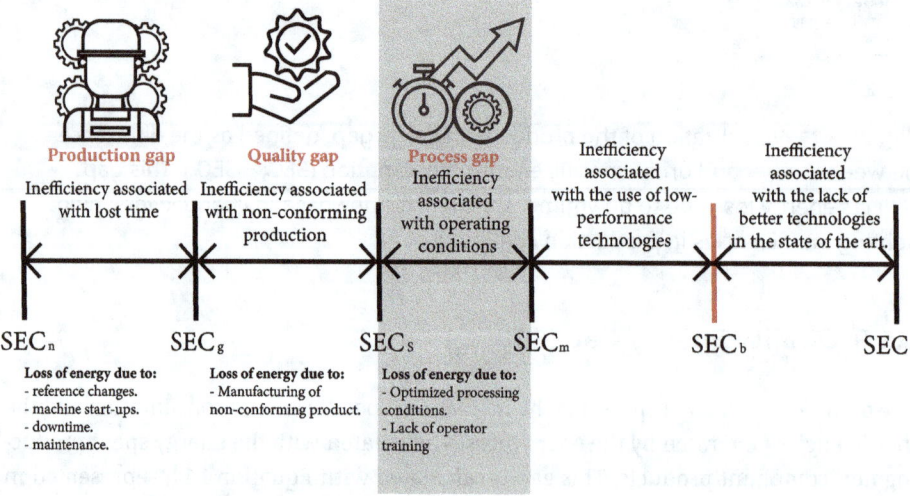

Figure 4.13 Representation of the process energy gap (SEC_s – SEC_m), indicating inefficiencies caused by operating outside optimal process parameters. Closing this gap requires process tuning and staff training

The process energy gap represents an increase in specific energy consumption and, therefore, the inefficiencies generated by operating the process at operating conditions far from the optimum conditions that maximize productivity and reduce specific energy consumption. This gap is calculated in Equation 4.15 and is presented in Figure 4.13.

$$EG_{process} = SEC_s - SEC_m \tag{4.15}$$

This is one of the most difficult energy gaps to determine since, usually, by the time it is known, it is because the necessary efforts have already been made to close it and, therefore, it no longer exists. By definition, the SEC_m is the value of SEC_s when the process parameters are modified in search of the best energy yield. For this reason, when SEC_m is defined, the energy efficiency of the production line is already improved and, therefore, $EG_{process}$ becomes zero.

4.1.11 Technology Energy Gap (EG$_{tech}$)

The technology energy gap represents the increase in specific energy consumption and, therefore, the inefficiencies generated by producing low-energy efficiency technologies. This gap is calculated with Equation 4.16 and is presented in Figure 4.14.

$$EG_{tech} = SEC_m - SEC_b \tag{4.16}$$

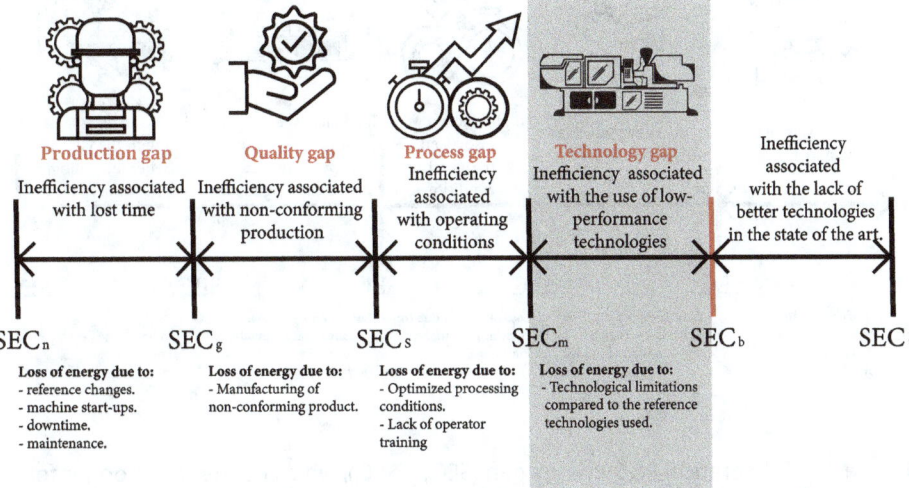

Figure 4.14 Diagram of the technology energy gap ($SEC_m - SEC_b$), highlighting the difference between current technological performance and that of best-available technologies. This gap supports investment decisions and technology benchmarking

The technology energy gap is comparative. If you want to determine the energy gap between the technology of interest and the best available technology on the market, then the value of SEC_b should be that of the leading energy-efficient technology in the current state of the art. If you want to determine why, if the plant has two equal machines, does one have better energy and production performance than the other, then the SEC_m value of the best-performing machine can be taken as the SEC_b of the second one. This comparison shows the technological gap between the two. Sometimes, these differences may be due to wear phenomena, difficulties in operating some equipment components, and undocumented technological changes. The value of the specific reference energy consumption will depend on the question to be answered, and its establishment is subject to the creativity and needs of the user.

4.1.12 Research & Development Energy Gap (EG$_{R\&D}$)

The R&D energy gap represents the opportunity to develop better technology in terms of energy efficiency. This gap is calculated with Equation 4.17 and is presented in Figure 4.15.

$$EG_{R\&D} = SEC_b - SEC_t \qquad\qquad (4.17)$$

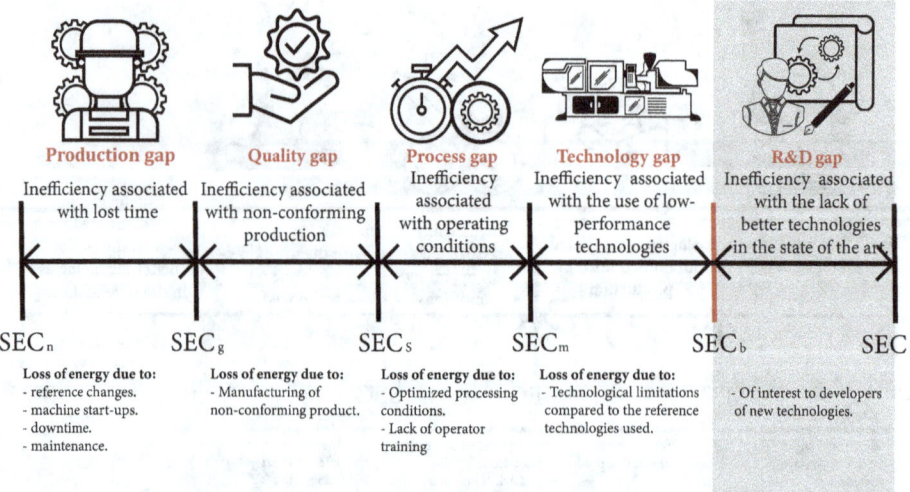

Figure 4.15 Theoretical R&D energy gap (SEC_b – SEC_t), which defines the frontier for innovation. It indicates the potential improvement in energy efficiency that could be achieved through new technology development

This gap definition only makes sense when the SEC_b is calculated using the most energy-efficient technology available in the state of the art. The definition of this gap is of great value to machine development companies as an indicator of technology energy efficiency improvement from the early stages of the production life cycle (R&D, conceptual design, detailed design, etc.).

4.1.13 Other Considerations for the Implementation of EGM

In the EGM, the second rule to be followed is the following:

$$SEC_n - SEC_t = EG_{production} + EG_{quality} + EG_{process} + EG_{tech} + EG_{R\&D} \qquad (4.18)$$

According to Equation 4.18, the total energy inefficiency comprises five sources of inefficiency through each of the described gaps. An intrinsic advantage of the EGM is that it allows evaluation of the performance of an EAC in all orders, using the energy gaps as a single common indicator. The magnitude of each of the energy gaps can be used to make decisions that guide the company to focus its improvement efforts, considering the following action principles of the method:

1. Establishing an action plan to close the most significant energy gap

2. If two energy gaps have similar magnitudes, priority should be given to actions that close the lower energy gap in the hierarchical order presented in Figure 4.15. Actions to address these gaps have the effect of helping to reduce the higher energy gaps in the hierarchical order, but this does not happen in the opposite direction. In other words, actions aimed at reducing the technology energy gap can generate reductions in the process, quality, and production energy gaps. Still, actions aimed at reducing some of these three will not mitigate the technology energy gap.

4.1.14 Energy Efficiency from SEC Definitions in the EGM

The absolute energy efficiency of the EAC can be defined in terms of the specific energy consumption values such as:

$$\eta_{E_{abs}} = \frac{SEC_t}{SEC_n} \cdot \frac{W_t}{W_c} = \frac{SEC_t}{SEC_n} \cdot \frac{1}{(1 - f_{non-compliant\ product})} \qquad (4.19)$$

When evaluating the efficiency of the EAC based on Equation 4.19, the results must be carefully interpreted, since low absolute energy efficiency values do not necessarily mean that the EAC is operating inefficiently. This is because no system can achieve 100% energy efficiency. Each EAC has a maximum energy efficiency that it can reach, which is inherent to the technology used in the process [7]. Improving it requires improving technology. This energy efficiency is given by:

$$\eta_{E_{max}} = \frac{SEC_t}{SEC_m} \qquad (4.20)$$

The energy efficiency that best represents the energy performance of the EAC during operation and use is the relative energy efficiency, which is defined as:

$$\eta_{Erel} = \frac{\eta_{Eabs}}{\eta_{Emax}} \cdot \frac{W_t}{W_c} = \frac{SEC_m}{SEC_n} \cdot \frac{1}{(1 - f_{non-compliant\ product})} \tag{4.21}$$

In the work of Estrada et al. [7], the energy efficiency of six different plasticizing units is evaluated: five different screw configurations in a grooved feed extruder (GFE) and one in a grooved plasticating extruder (GPE). Each configuration was evaluated with the same homogenizing and mixing zone geometry, the same temperature profile, the same polymer (PP MFI 5 g/10 min at 230 °C and a load of 2.16 kg), and nine different operating conditions corresponding to three screw rotation speeds and three screw restriction levels. An extensive discussion of the energy performance of each of the evaluated configurations is made in the paper. Energy efficiency is presented in Table 4.4.

Regarding the results presented in Table 4.4, the absolute energy efficiency of all plasticizing unit configurations is very low and is in the range 49–59%. However, when the relative energy efficiency is evaluated, it rises to 61–86% due to the maximum energy efficiency, reaching values of 62–82% for the case studied; these values indicate that there is a better use of the capabilities of the technology than the absolute energy efficiency predicts. In this paper, $SEC_m = a$, where a is the slope of the average demand curve in steady-state operation vs. mass flow. The SEC_s has an asymptotic curve at a minimum value which is precisely the value of a. This implies that SEC_s will never reach this value, but it is a good approximation to SEC_m in the absence of better information, as described in previous chapters.

Table 4.4 η_{Eabs}, η_{Emax} and η_{Erel} Values Obtained for the Different Plasticization Unit Configurations Evaluated in [7]

PUT-screw	Die restriction	η_{Eabs} (%)	η_{Emax} (%)	η_{Erel} (%)	$a = SEC_{min}$	b	R^2
GFE-Screw 1	Low	49.50	68.50	72.26	0.257	5.745	0.9837
	Medium	50.10	67.50	74.22	0.261	5.293	0.9987
	Very high	48.40	58.90	82.17	0.294	5.265	0.9980
GFE-Screw 2	Low	47.80	63.60	75.16	0.272	4.834	0.9939
	Medium	48.10	62.30	77.21	0.279	4.897	0.9912
	Very high	46.70	54.20	86.16	0.312	4.333	0.9921
GFE-Screw 3	Low	52.20	73.30	71.21	0.227	5.269	0.9991
	Medium	51.00	74.10	68.83	0.226	5.947	0.9956
	Very high	–	69.80	–	0.243	6.081	0.9909

PUT-screw	Die restriction	$\eta_{E_{abs}}$ (%)	$\eta_{E_{max}}$ (%)	$\eta_{E_{rel}}$ (%)	$a = SEC_{min}$	b	R^2
GFE-Screw 4	Low	50.60	71.30	70.97	0.239	5.036	0.9952
	Medium	49.80	68.40	72.81	0.251	5.261	0.9974
	Very high	–	67.90	–	0.253	5.596	0.9998
GFE-Screw 5	Low	51.60	84.20	61.28	0.202	7.397	0.9883
	Medium	50.10	81.10	61.78	0.213	7.421	0.9953
	Very high	49.40	73.80	66.94	0.235	7.118	0.9923
GPE-Helibar	Low	59.60	77.80	76.61	0.175	7.210	0.9813
	Medium	58.30	77.00	75.71	0.179	7.526	0.9782
	Very high	57.80	77.90	74.20	0.175	7.629	0.9970

4.2 Other Diagnostic Tools of the EGM

The proper deployment of the energy gaps method starts from the effective determination of the net specific energy consumption (SEC_n) and the stable specific energy consumption (SEC_s). The behavior of the energy gaps in the EAC depends not only on the definition of the EAC but also on the specific product being produced. A simple EGM analysis requires selection of a target product to track the behavior of energy gaps and take action to improve its energy performance.

In many instances, the diagnosis of the EAC's performance with one product leads to actions that improve the EAC's energy performance with all or most of the products produced in it. This is further facilitated when the EAC produces few products. In these cases, an approximation of the SEC_n determination by M&T, or simply determining it from the total consumption and compliant production during the production of several batches of the same product, may be sufficient. To establish SEC_s, demand, and the corresponding mass flow, are tracked under steady-state production conditions. But what happens when the EAC produces many different products, and it is complex to select the products that can most impact the energy performance of the EAC?

The EGM proposes a new way of monitoring the SEC_n and SEC_s. It is inspired by the tools of monitoring and control by improvement objectives to determine the PCL, ABT, and PCC, which represent the behavior of the EAC. Also, these tools allow identification of aspects needing improvement or the decisions required to improve EAC energy performance and its energy and productive efficiency. This transforms the M&T tools into an essential diagnostic tool. To review the definition of these concepts in detail, we recommend referring to Chapter 3.

This analysis requires more detailed information on energy consumption and production since it is performed by production orders and not by regular periods. Each production order is associated, at least, with a quantity of compliant products produced, a product reference, a material, a mold, and a machine with which it is manufactured. In addition, the start and end date and time of the production order (PO) are usually reported. Finding this information in companies is becoming easier with the progress, availability, and democratization of information systems. On the other hand, it is necessary to determine the energy consumption during the period corresponding to the production of the PO.

In the era of Industry 4.0, the offer of energy consumption monitoring systems in plastics production plants is varied, easy to implement, and inexpensive. However, in the absence of this possibility and for diagnostic purposes, it is sufficient to connect the EAC to an energy consumption meter or a demand monitor for the time necessary for enough production orders to run in the EAC to enrich the analysis.

Consolidating the necessary energy information for each reference, mold, or machine during the analysis period can be a complex task due to the amount of data to handle. In this sense, it is necessary to develop automated calculation routines to obtain the required values.

4.2.1 Determination of SEC$_n$ Using Diagnostic Tools

In the following, the diagnostic tools developed by the EGM will be presented, which are inspired by M&T tools to achieve a more effective diagnosis of the EAC and provide a way to establish the appropriate value of SEC$_n$ for energy gap analysis.

4.2.1.1 Determination of Performance Characteristic Line for Diagnostic (PCLD) Purposes

To determine the behavior of SEC$_n$, the first thing to obtain is the PCL from a diagram of energy consumption vs. compliant production, as would be done in the M&T method. The difference is that each point in the diagram corresponds to the consolidation of energy consumption and compliant production, according to some criterion related to production orders, and not a regular period. The criteria can be by machine, mold, or reference. The analysis is not performed by production order because production orders are not repeated, and therefore, the conclusions would not be helpful. The period selected to perform the analysis must be broad enough to have at least 12 points on the diagram for the selected criterion, and the coefficient of determination to accept that there is an adequate correlation between the line equation and the actual points must be at least 0.7 ($R^2 > 0.7$).

The interpretation of the PCL is the same as that of the M&T, for which the reader is recommended to review Chapter 3. The slope of the line is the variable load, while the intercept point represents the base load. However, the base load represents energy consumption per change of reference, mold, or machine, depending on the criteria chosen to obtain the line. As a result of the exercise, a line configuration is obtained that allows the establishment of the performance of the EAC in terms of energy consumption. The molds, references, or machines corresponding to the points below the line have a better performance than the average, whereas those above are those have a poorer performance. In this way, the PCL becomes a diagnostic tool, which is why it has been called a "performance characteristic line for diagnostic" (PCLD) in EGM.

4.2.1.2 Determination of the Activity-Based Target from Diagnostic (ABTD) Data

The PCLD represents the average behavior of the consumption vs. production data according to the selected criteria. This implies that there are data above and below the PCLD. As previously established, the points above the average indicate lower energy performance, and the points below indicate better use of energy resources. It is then possible to perform a new regression analysis with the latter points (those that are better than the PCLD), which provides a target or result that would be challenging but achievable if actions are taken to promote higher energy performance of the points that are above the PCLD, as shown in Figure 4.16. This line allows calculation of the expected improvements and savings from the intervention on the EAC if the actions on the underperforming points (higher than PCLD) are effective.

Figure 4.16 ABTD derived from points showing better-than-average energy performance. This line represents a realistic yet challenging goal for reducing energy consumption in underperforming units

4.2.1.3 Determination of the Performance Characteristic Curve for Diagnostic Purposes (PCCD)

As in M&T, the PCLD is used to obtain the equation for the PCCD, and this behavior is represented in a diagram of SEC_n vs. compliant production, as shown in Figure 4.17. Two lines are identified on the graph: the PCLD and a horizontal line corresponding to the average net specific consumption for the period under analysis ($SEC_{n_{average}}$). The value is calculated as follows, where n is the number of references, molds or machines analyzed:

$$SEC_{n_{average}} = \frac{\sum_{i}^{n} Energy_{consumption_i}}{\sum_{i}^{n} Production_{compliant_i}} \tag{4.22}$$

All references, molds or machines whose performance is above the $SEC_{n_{average}}$ line contribute to the increase of this average specific energy consumption. Conversely, all points whose performance is below the $SEC_{n_{average}}$ line contribute to the improvement (reduction) of the average specific energy consumption. The intersection between the $SEC_{n_{average}}$ line and the PCCD defines a compliant production value called CP*. This value represents the minimum lot size, according to the selected criterion, from which energy efficiencies in the EAC allow to equal or improve the $SEC_{n_{average}}$.

On the other hand, the points (references, molds, or machines) above the PCCD exhibit poor energy performance compared to the average production behavior, while those below have superior energy performance. Correctly differentiating between energy efficiency and energy performance is crucial to direct the actions required for improvement.

Figure 4.17 provides further insight into the concept of energy efficiency and performance. Understanding that the points in this graph represent references, molds, or machines, it can be said that points A and B have the same energy performance because they are above the PCLD, but point B offers higher energy efficiency. Points A and C have the same energy efficiency, but point C has poorer energy performance because it is above the PCLD. An analysis only considering SEC_n would prioritize the intervention of point D over point C, as it has a lower energy efficiency. However, improving the energy efficiency of point D is just a matter of scheduling a larger batch. In contrast, improving the energy performance of point C requires a planned engineering intervention.

To expand on this concept, four well-defined areas are presented in Figure 4.17. The red area above the $SEC_{n_{average}}$ line and above the PCCD is labeled A_{BPBE}, where points exhibit poor energy performance and low energy efficiency. These are critical points whose improvement depends on radical engineering or production decisions, and their improvement is usually associated with a high cost. The yellow area A_{GPBE}, which is above the $SEC_{n_{average}}$ value and below the PCCD, corresponds to points that have good energy performance but low energy efficiency. To improve the points in A_{BFGE} only requires increasing the production batch size, preferably above CP*. The points in the blue area A_{BPGE}, which is above the PCCD and below the $SEC_{n_{average}}$ value, are not a priority. It is very likely that during continuous improvement, there will come a time when a new diagnosis makes them points that merit intervention. The green area, called A_{GPGE}, contains the points of good energy performance and good energy efficiency.

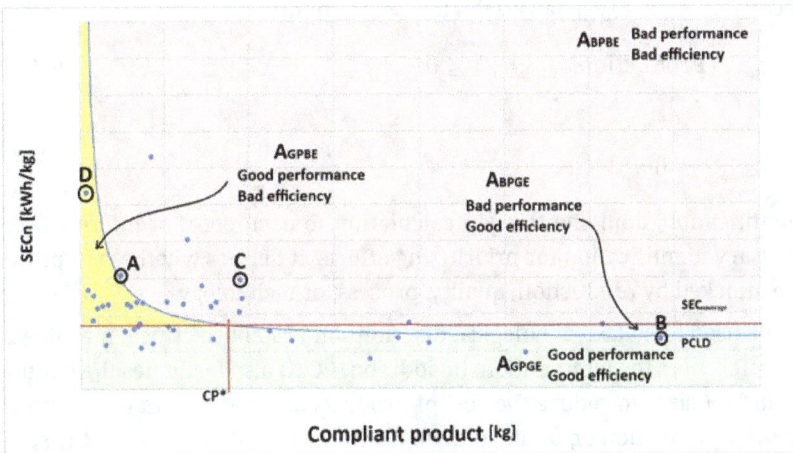

Figure 4.17 PCCD, plotting SEC_n vs. compliant production. The curve differentiates between energy efficiency and energy performance to prioritize corrective actions. This figure identifies four areas and the actions for improving each point (mold, reference or machine) with different ones according to the area in which it is located

Considering that SEC_{n_i} is the actual net specific energy consumption of point i, $SEC_{n_{PCCD}}(Production_{compliant_i})$ is the specific energy consumption evaluated through the PCCD at the compliant production value of point i, and $SEC_{n_{average}}$ is the average specific energy consumption of the period under analysis, the points located in area A_{BPBE} meet the following condition:

$$SEC_{n_i} - SEC_{n_{PCCD}}(Production_{compliant_i}) > 0 \qquad (4.23)$$

and

$$SEC_{n_i} - SEC_{n_{average}} > 0$$

The points located in area A_{GPBE} satisfy the following conditions:

$$SEC_{n_i} - SEC_{n_{PCCD}}(Production_{compliant_i}) \leq 0 \qquad (4.24)$$

and

$$SEC_{n_i} - SEC_{n_{average}} > 0$$

The points located in area A_{BFGE} meet the following conditions:

$$SEC_{n_i} - SEC_{n_{PCCD}}(\text{Production}_{\text{compliant}_i}) > 0 \tag{4.25}$$

and

$$SEC_{n_i} - SEC_{n_{average}} \leq 0$$

The points located in area A_{GPGE} satisfy the following conditions:

$$SEC_{n_i} - SEC_{n_{PCCD}}(\text{Production}_{\text{compliant}_i}) \leq 0 \tag{4.26}$$

and

$$SEC_{n_i} - SEC_{n_{average}} \leq 0$$

From this identification, applying the gap calculation to each point of interest provides the necessary identification for prioritizing efforts. It defines whether the problem should be attacked by production, quality, process, or technology.

The PCCD can be used to make technological decisions and establish critical machines, molds, and/or references for intervention. In addition, PCCD also facilitates the definition of minimum lot sizes to reduce the cost of products associated with energy consumption for their production or, failing that, to transfer the cost of energy to the specific product that generates it.

4.2.2 Determination of SEC_s Using Diagnostic Tools

In the same way that PCLD and PCCD are constructed for the determination of SEC_n, an analogous analysis can be performed for SEC_s, a parameter that offers other equally important diagnostic elements. The difference is that SEC_s is mass-flow-dependent. Therefore, what is plotted is the average demand under steady-state operating conditions vs. mass flow. The equation for the energy performance characteristic line under steady-state conditions has the following form:

$$\text{AVGPower}_{\text{demand}}[\text{kW}] = a \cdot \dot{m}\left[\frac{\text{kg}}{\text{h}}\right] + b \tag{4.27}$$

where:

a is the variable load [kWh/kg]

b is the fixed demand [kW].

The latter represents the average demand of the EAC while waiting for a change of reference, mold, or machine. To accept the correlation as good, the coefficient of determination should be greater than 0.7. Although the information also allows the determination of an ABTD line for the SEC_s, this is of little use. However, the elaboration of the PCCD, as shown in Figure 4.18, is significant.

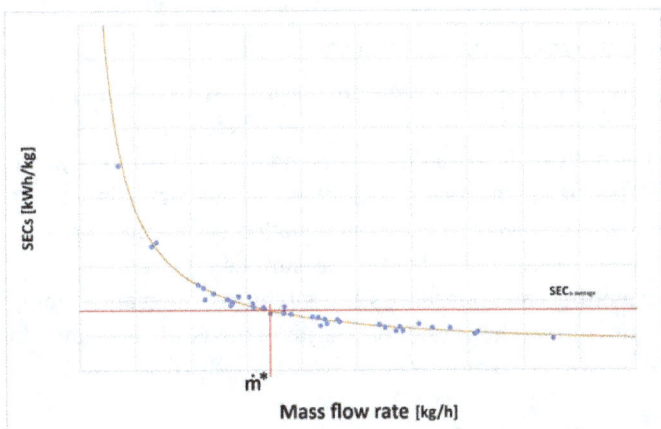

Figure 4.18 PCCD focused on the stable specific energy consumption (SEC_s), enabling identification of minimum mass flow rates for optimal performance and guiding equipment selection and minimum mass flow rate target

First, with the equations describing the SEC_n and SEC_s behavior of the EAC, and if an estimated SEC_b is available for the type of technology used, sufficient information is available to define two combined energy gaps: the production + quality energy gap ($SEC_n - SEC_s$) and the process + technology energy gap ($SEC_s - SEC_b$). In this way, it is possible to establish how the EAC is expected to behave energetically from the moment the production order is assembled in the production program.

On the other hand, it is possible to determine the minimum mass flow ($\dot{m}^*$) at which the EAC must work, as shown in Figure 4.18. If a horizontal line is drawn at the value of SEC_s corresponding to $SEC_{n average}$, it will cross the PCCD at $\dot{m}^*$. It is worth clarifying that at this point $SEC_n - SEC_s = 0$, and therefore, the value of $\dot{m}^*$ is the value of mass flow required to reach a stable specific energy consumption value. This point does not allow downtime or energy expenditure in the production of non-compliant products, so as not to increase the average specific energy consumption of the period under analysis.

For all the above reasons, the determination of the PCCD for SEC_n and for SEC_s has high value for making decisions on production scheduling, product costing, and even design and specification of references, molds, and machinery in a polymer processing plant.

4.2.3 Diagnosis with Determination of PCLD, ABTD and PCCD

To exemplify the use of the tools presented, a couple of cases will be reported that may help to understand the implementation of the method better.

4.2.3.1 EAC: EPDM Rubber Profile Extrusion Line

A company dedicated to manufacturing vulcanized rubber profiles wanted, in principle, to compare the performance of one of its extrusion lines when it operates using a melt pump and when it does not use it. To do this, a demand data acquisition system was connected to the line's totalizer, which recorded information every minute. The data acquisition equipment was connected for 7 days in both cases. In each case, 12 and 15 product references were produced during the measurement period. An analysis was then performed by plotting each reference's energy consumption and compliant production during the measurement period to determine the PCLD, as shown in Figure 4.19.

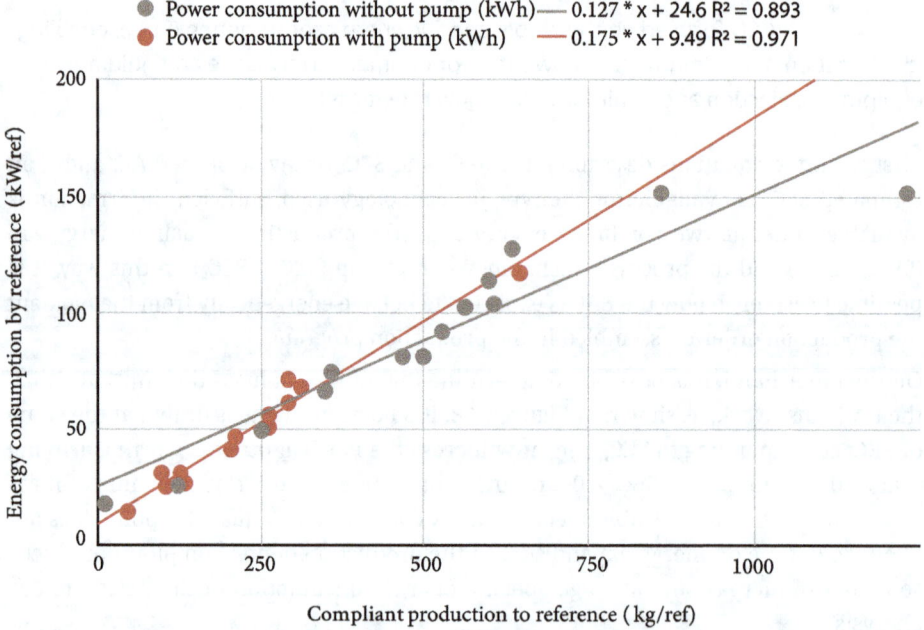

Figure 4.19 PCLD comparing extrusion with a melt pump (red) and without a melt pump (brown) for a vulcanized rubber profile extruder. This highlights differences in base and variable loads, supporting decisions on when to use melt pumps for energy optimization

The trend lines obtained are very representative of the process, as they have a correlation coefficient higher than 0.7. This trend line is given by:

$$\text{Energy}_{\text{consumption}} \; [\text{kWh}] = a \cdot \text{Production}_{\text{compliant}} \; [\text{kg}] + b \qquad (4.28)$$

where:

a is the variable load [kWh/kg]

b is the base load [kWh/ref].

As shown in Figure 4.19, when working with a pump, there is a lower base load (9.49 kWh/ref vs. 24.6 kWh/ref) but a larger variable load (0.175 kWh/kg vs. 0.127 kWh/kg). Although the production rates are not comparable in this example, the lower base load implies that the melt pump reduces energy consumption by making the system available to operate during non-productive times. This occurs because it facilitates tuning, reaching stability faster during reference changes. On the other hand, the higher variable load indicates reduced line efficiency with the presence of the melt pump. This makes sense since, under these conditions, the line has an additional consumer for similar production levels.

However, the best tool to analyze performance of the line is the PCCD or the SEC_n vs. $\text{Production}_{\text{compliant}}$ diagram, as shown in Figure 4.20.

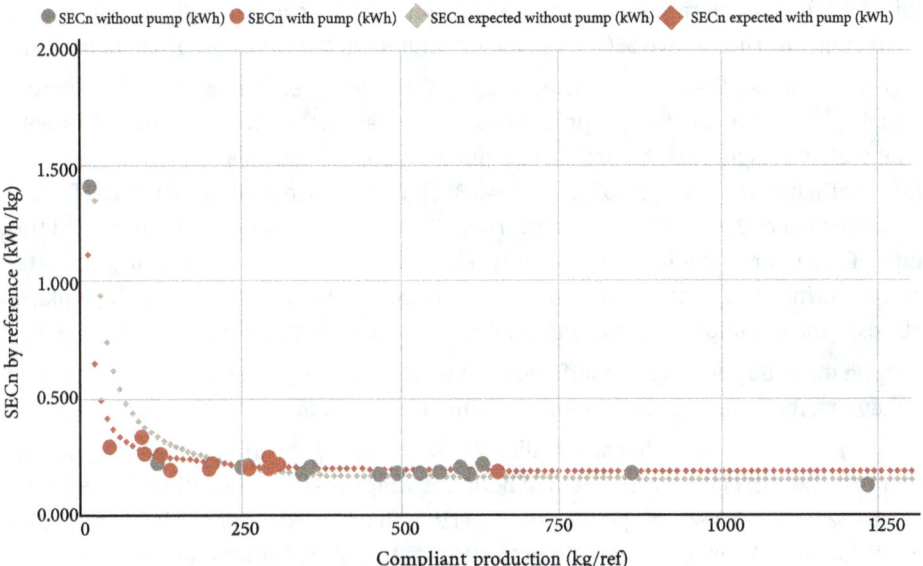

Figure 4.20 Diagram of SEC_n vs. $\text{Production}_{\text{compliant}}$ for a vulcanized rubber profile extruder with a melt pump (red) and without melt pump (brown). This shows how the melt pump benefits short runs but may increase SEC_n in longer batches, informing minimum batch size planning

The PCCD is obtained from the PCL using the following equation:

$$SEC_n \left[\frac{kWh}{kg} \right] = a + \frac{b}{Production_{compliant}[kg]} \qquad (4.29)$$

This trend line is essential in predicting how the production line will behave regarding net-specific energy consumption as compliant production increases.

In Figure 4.20, it can be seen that, both with and without a melt pump, the efficiency of the production line increases with increasing conformal production, as SEC_n decreases with compliant production (the base load is amortized over greater production). On the other hand, the production line's performance is superior when working with a melt pump only when the batch size is less than approximately 300 kg. For larger batch sizes, the performance of the extrusion line without a melt pump is superior, since for the same level of compliant production, the SEC_n is lower, and consequently, the energy efficiency is higher. This is precisely the meaning of better performance.

Another use of Figure 4.20 is determining the minimum recommended batch size. When batch sizes are very small, the SEC_n increases dramatically (the base load is significant). In the case of the production line without a pump, the minimum batch size would be approximately 250 kg/ref, while with the pump, it could be approximately 125 kg/ref. In this case, the melt pump provides more flexibility for the company when dealing with short runs. For runs above 250 kg/ref, the investment in the melt pump is not justified.

Figure 4.21 shows the energy performance of the vulcanized rubber extruder without a melt pump as a function of process speed or mass flow. The data correlates very well with a straight line, according to Equation 4.28. The graph shows a high correlation coefficient for this type of case ($R^2 = 0.862$), with a base load of 6.01 kW/ref and a variable load of 0.0754 kWh/kg. The variable load is very low but consistent with this type of extruder. In rubber extrusion processes, the screw-cylinder assembly operates as a metering device for the mixture. The mixture is not heated significantly or plasticized, so the enthalpy change is minimal, as there is no state change.

Despite the good correlation coefficient in Figure 4.21, it is possible to see two marked behaviors. By separating these behaviors into two trend lines, Figure 4.22 is obtained.

Figure 4.22 shows two almost parallel PCLD curves with different energy performances. The correlation coefficient of both lines improves significantly ($R^2 = 0.979$ and $R^2 = 0.882$ vs. $R^2 = 0.862$), supporting the assertion that the extruder exhibits two different behaviors. After reviewing the production information, it was found that this behavior was due to the existence of two different rubber formulas used to obtain the products: one formula called a "technical formula", which corresponds to the PCLD characterized as formula 1, and the other, called an "economic formula", which corresponds to the PCLD characterized as formula 2. The technical formula has a higher cost but superior energy performance compared to the economic formula.

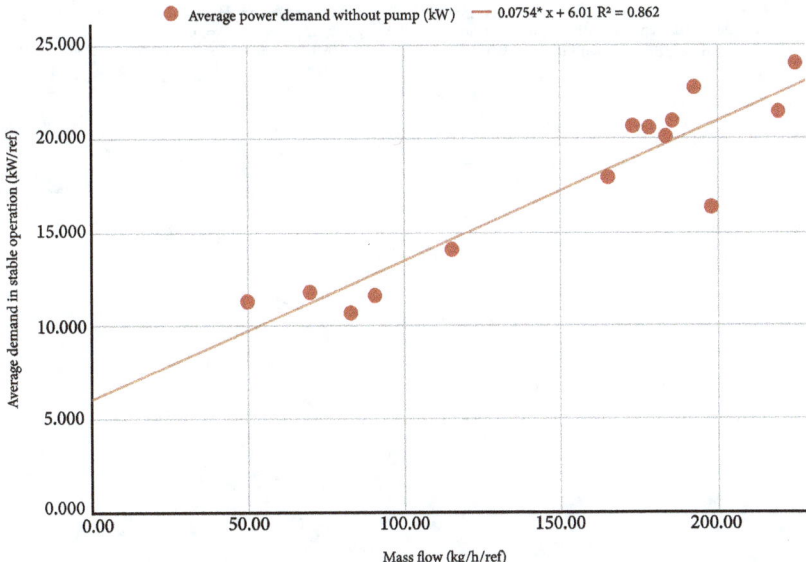

Figure 4.21 Average energy demand as a function of mass flow in a vulcanized rubber extrusion line operating without a melt pump. The figure reveals linear behavior with distinct operational regimes, which is useful for establishing steady-state energy consumption baselines

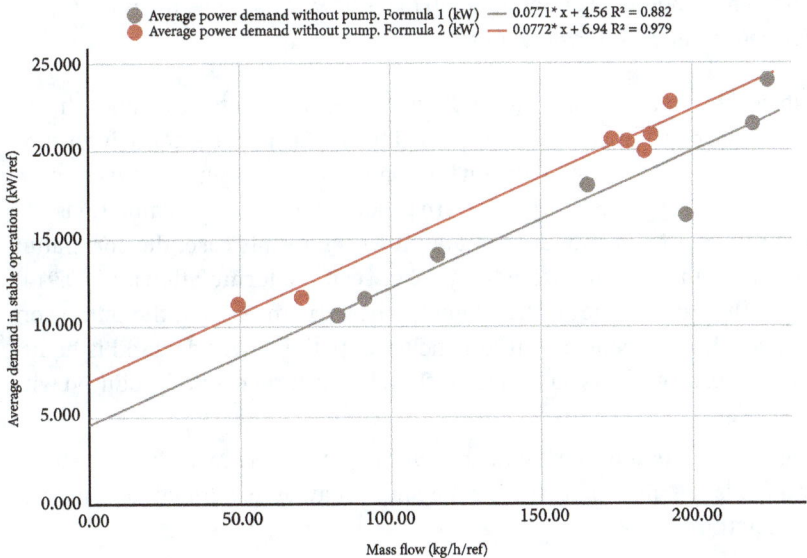

Figure 4.22 Comparison of energy demand under steady-state conditions for the vulcanized rubber extruder for two formulations – economic (brown) and technical (red) – processed without a melt pump. The plot demonstrates the superior energy performance of the technical formulation, with consistent separation in demand curves

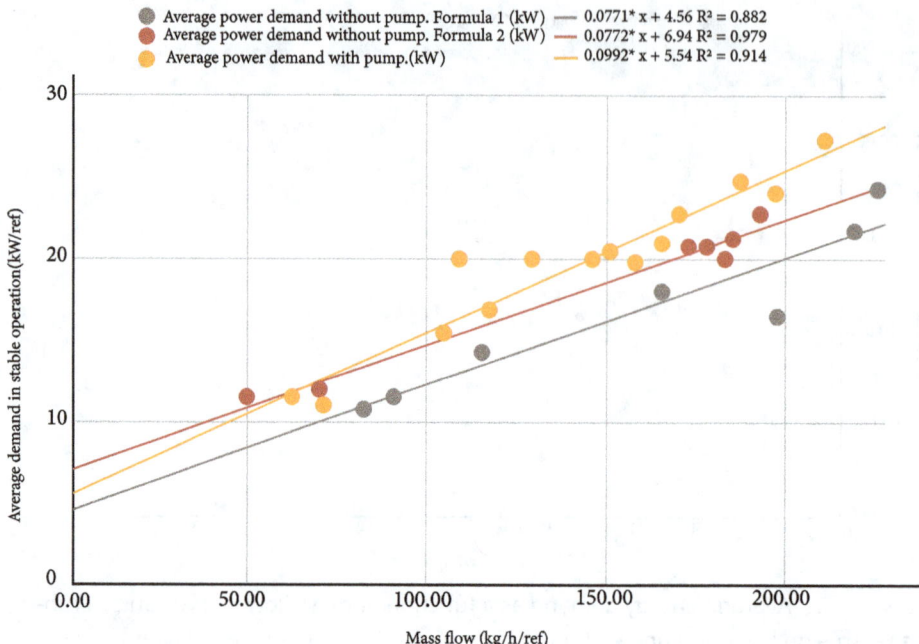

Figure 4.23 Average power demand in steady-state operation for a vulcanized rubber extruder under three configurations: with melt pump (yellow) and without melt pump using two different formulations (brown and red). The figure shows increased energy demand with the melt pump, especially at high mass flow rates, while maintaining similar performance between formulations

Figure 4.23 shows the behavior of the extruder when the melt pump is coupled. In this case, the variable load is higher and can be explained by the fact that there is an additional consumer connected to the extrusion line, namely the melt pump or gear pump. Among the measured references, profiles were produced with both formulations: the technical formula and the economic formula. However, in this case, the correlation coefficient is very high without differentiating between the formulations ($R^2 = 0.914$), which indicates that they do not differ in energy performance when the pump controls the extruder flow. Moreover, with the melt pump, the power demand is higher when the process speed exceeds 60 kg/h mass flow. Few references are produced with lower process speeds.

From the linear correlation of the average power data with the mass flow, in steady state operation, it is possible to determine the SEC$_s$ behavior as a function of $\dot{m}$ from the following equation:

$$\text{SEC}_s \left[\frac{\text{kWh}}{\text{kg}} \right] = a \left[\frac{\text{kWh}}{\text{kg}} \right] + \frac{b\,[\text{kW}]}{\dot{m} \left[\frac{\text{kg}}{\text{h}} \right]} \tag{4.30}$$

The result of Equation 4.30 is plotted in Figure 4.24, from which the following conclusions can be drawn:

- The economic formula (formula 2) has a higher specific energy consumption under stable operating conditions (SEC$_s$) than the technical formula when operating without a melt pump. For example, at 50 kg/h, the difference is approximately 0.048 kWh/kg. This difference is considerably reduced at 250 kg/h when the difference is 0.001 kWh/kg. Both curves are asymptotic at a value of $a = 0.077$ kWh/kg, which is, theoretically, the minimum value of SEC$_s$ that is possible to obtain from the technology when $\dot{m} \rightarrow \infty$.

- In comparison, the energy performance of the extruder with a melt pump at high and low process speeds is lower than that of the extruder without the pump. At 50 kg/h, it is 0.042 kWh/kg, while at 250 kg/h, it is 0.026 kWh/kg. The minimum SEC$_s$ value that is possible to have in the vulcanized rubber extruder studied with the melt pump coupling is $a = 0.0992$ kWh/kg, which is 0.0221 kWh/kg higher than in the extruder without the melt pump. This is not necessarily detrimental if the melt pump allows a reduction in non-compliant product or the downtimes due to reference change to achieve a lower SEC$_s$ value. Otherwise, its use is not justified.

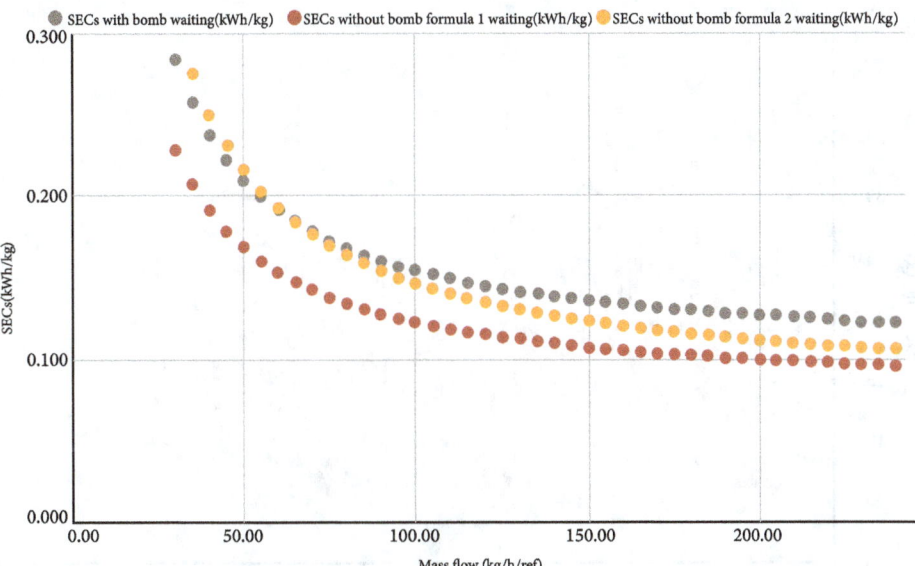

Figure 4.24 Plot of stable specific energy consumption (SEC$_s$) as a function of mass flow for a vulcanized rubber extruder with a melt pump (brown) and without a melt pump for formula 1 (red) and formula 2 (yellow). The asymptotic behavior illustrates theoretical energy efficiency limits and highlights the impact of formulation and equipment setup on energy performance

4.2.3.2 EAC: Thermoplastic Injection Line

A company dedicated to the manufacture of thermoplastic injection- molded household products wanted to establish criteria to determine the size of the minimum production batches and the cost of the products based on the energy performance of the process. Additionally, it wanted to identify actions to improve the energy performance of the line.

A hydraulic injection molding machine with 200 t of clamping and a capacity of 291 g (measured with PS) was selected as the EAC. The analysis was conducted on 299 production orders corresponding to 105 references produced in 43 different molds. The analysis period corresponded to 4 months of production (note that this is not a seasonal analysis). The EAC was set up to obtain demand measurements every 30 s, and an information system was used that allowed determination for each production order of the following: start time, end time, product reference, mold used, compliant production, non-compliant production, downtime, part weight, and number of mold cavities.

The energy monitoring system was not integrated with the information system. Still, it was easy to connect the consumption with the production order by the start time and end time of the PO. In this way, it was possible to obtain demand graphs for each production order, as shown in Figure 4.25.

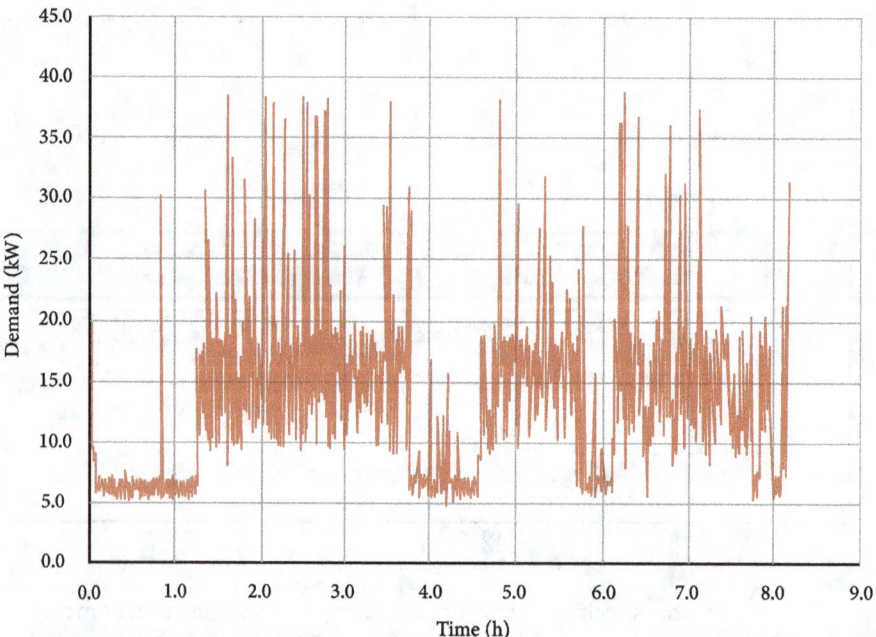

Figure 4.25 Real-time energy demand profile during production order 888961, working with mold MOL4143 and part number 2-1004817 for injection molding. The figure shows fluctuations due to process transitions and setup activities, serving as a baseline for stability and consumption analysis

Due to the frequency of the demand information, a lot of noise could have made the analysis difficult. For this purpose, a noise attenuation and reduction routine was applied. Subsequently, the curve was integrated using Simpson's rule to obtain consumption information, and the average demand under stable production conditions was established through a routine developed for this purpose. This way, the information presented in Figure 4.26 is obtained for each of the 299 production orders.

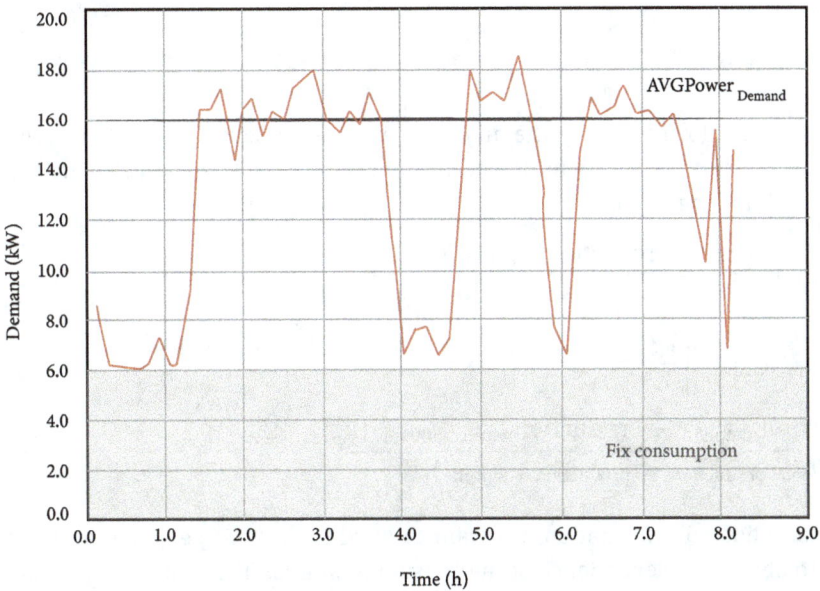

Figure 4.26 Filtered energy demand signal showing steady-state behavior and highlighting base load and average demand for PO 888961, working with mold MOL4143 and reference 2-1004817, attenuated to 1/20. This includes determining fixed consumption and average demand under steady-state production conditions

Table 4.5 Consumption and Production Information for the EAC under Study

Parameter (values related to the period stated)	Value	Unit
Start date of analysis	08/08/2023 23:22	—
End date of analysis	24/01/2024 14:00	—
Total consumption	36217.7	kWh
Total production	41,824.89	kg
Compliant production	40,689.47	kg
Non-compliant production	1,135.42	kg
Total production time	3,738.13	h

Table 4.5 Consumption and Production Information for the EAC under Study (*continued*)

Parameter (values related to the period stated)	Value	Unit
Total downtime	817.25	h
Effective production time	2,920.88	h
Average actual productivity	10.88	kg/h
Average effective productivity	14.32	kg/h
SEC$_n$	0.890	kWh/kg

The consolidated information for the analysis period is presented in Table 4.5. The effective and actual productivity for the period and each of the POs was calculated as shown in the following expressions:

$$Real_{productivity} \left[\frac{kg}{h}\right] = \frac{Production_{compliant}\,[kg]}{t_{total\;production}\,[h]} \tag{4.31}$$

$$Effective_{productivity} \left[\frac{kg}{h}\right] =$$

$$\frac{(Production_{compliant} - Production_{non-compliant})\,[kg]}{(t_{total\;production} - t_{production\;downtime})\,[h]} \tag{4.32}$$

With the consolidated information, we want to obtain the PCLD per mold and per reference to obtain the dependence of the consumption upon the compliant production. The results are shown in Figure 4.27 and Figure 4.28, respectively.

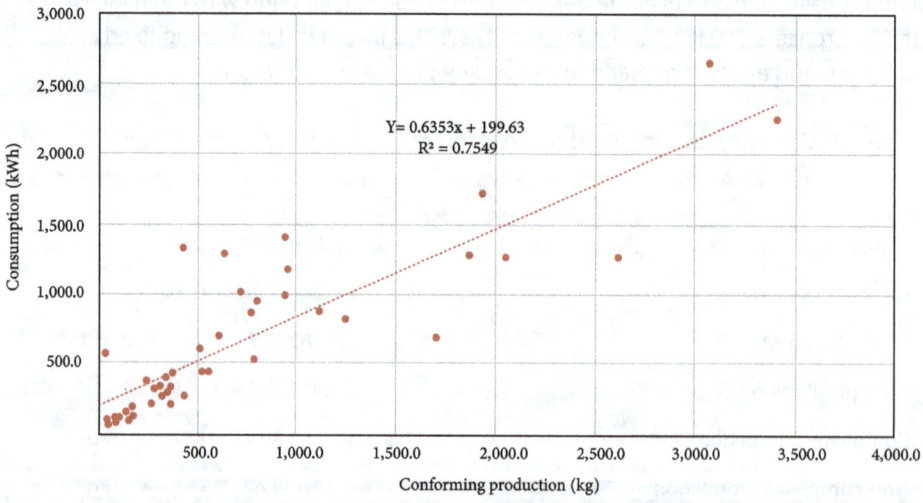

Figure 4.27 PCLD showing the relationship between compliant production and energy consumption by injection mold. The line reveals molds with above-average energy performance for prioritizing energy efficiency actions

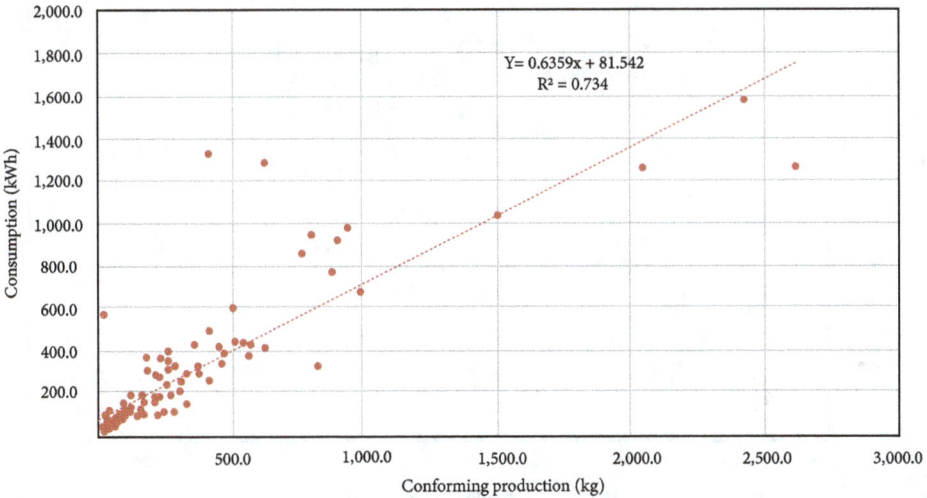

Figure 4.28 PCLD showing the relationship between compliant production and energy consumption by product reference. The line reveals references with below-average energy performance for prioritizing energy efficiency actions

In both cases, the coefficient of determination is adequate and establishes that the consumption information can be correlated with the compliant production through a linear regression. Additionally, the base load of the PCLD in Figure 4.27 has a value of 199.6 kWh per mold assembled, while the PCLD in Figure 4.28 has a base load of 81.5 kWh per product reference produced. These values are congruent, since about 2.4 times more references are assembled than molds. In other words, slightly more than two references are made on average in each mold. The variable load is very similar (0.6353 kWh/kg vs. 0.6359 kWh/kg, respectively). It should not change, since the efficiency with which the EAC consumes energy does not depend on the references or the molds if the production composition does not change. In both cases, the production composition is the same.

Figure 4.29 shows the determination of the PCLD purposes and the ABTD data for analysis by mold. The PCLD was determined with all the points in the graph, while the ABTD was obtained by correlating only the points below the PCLD and located in the shaded region.

The usual injector performance model from the PCLD is presented in the following equation:

$$\text{Energy}_{\text{consumption}_{\text{PCLD}}} \, [\text{kWh}] = 0.635 \cdot \text{Production}_{\text{compliant}} \, [\text{kg}] + 199.6 \qquad (4.33)$$

The correlation coefficient ($R^2 = 0.755$) is acceptable, and allows us to be confident that the model represents the behavior of the data. According to the model, the fixed consumption of the injection molding machine is 199.6 kWh per mold mounted on the machine, while the variable consumption is 0.635 kWh/kg produced.

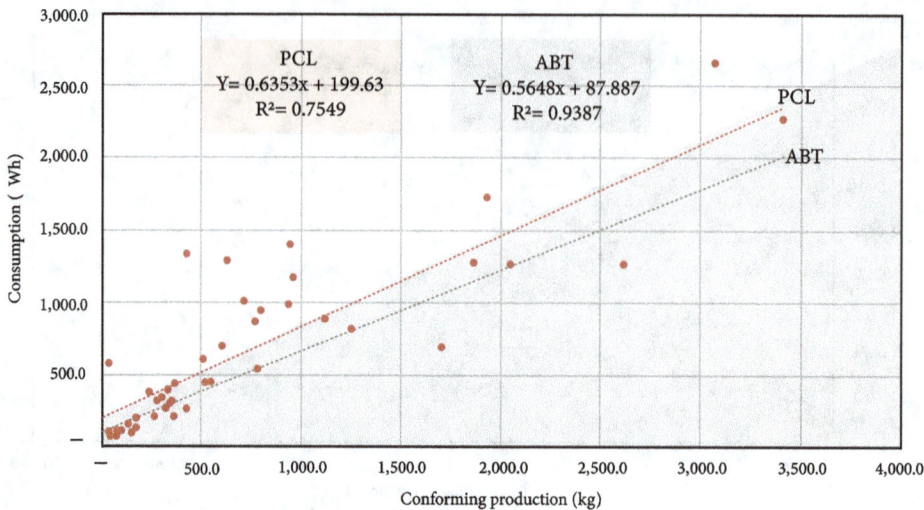

Figure 4.29 ABTD compared with the actual PCLD for energy consumption by mold. The gap between both curves quantifies improvement potential through process optimization and batch size adjustments

On the other hand, the ABTD curve has a much better coefficient of determination ($R^2 = 0.939$) and according to it, the consumption is represented by the following equation:

$$Energy_{consumption_{ABTD}} \ [\text{kWh}] = 0.565 \cdot Production_{compliant} \ [\text{kg}] + 87.9 \qquad (4.34)$$

If the actions for improving the energy performance of the injection molding machine are undertaken and a PCLD with behavior like the ABTD is achieved, the base load would drop from 199.6 kWh to 87.9 kWh per mold assembly. This means a reduction of 56.0%, and the variable load would drop from 0.635 kWh/kg to 0.565 kWh/kg, representing a reduction in energy consumption above the base load of 11.0% and, thus, an increase in EAC efficiency. Production must intervene in the molds corresponding to the points above the PCLD to achieve the target. The ABTD line is the target for the future PCLD line, and when the target is reached, a new ABTD line can be established. In a well-managed system, the gap between PCLD and ABTD gradually closes, making it increasingly difficult and costly to reach new targets. There will come a time when no further improvement is possible without significant intervention in EAC technology.

To determine which molds should be intervened and how, the PCCD analysis is performed. For this purpose, the total consumption is related to the total compliant production of the analysis period in the EAC, and a $SEC_{n_{global}} = 0.890$ kWH/kg (see Table 4.5).

Figure 4.30 presents the PCCD analysis that describes the behavior of the SEC_n for the analysis by molds. The PCCD by reference analysis is presented in Figure 4.31.

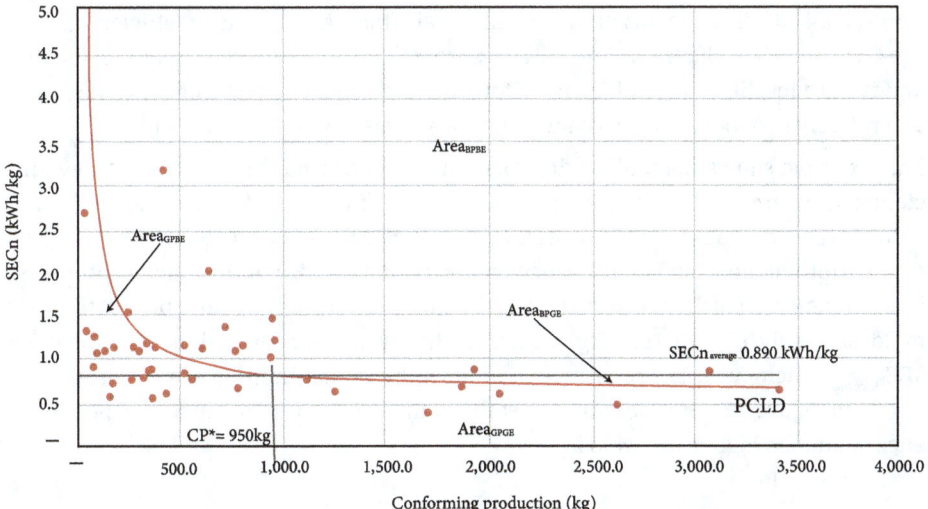

Figure 4.30 PCCD analysis of the SEC$_n$ from PCLD by molds. The curve supports identification of inefficient molds and guides decision-making on minimum production batch sizes to enhance energy performance

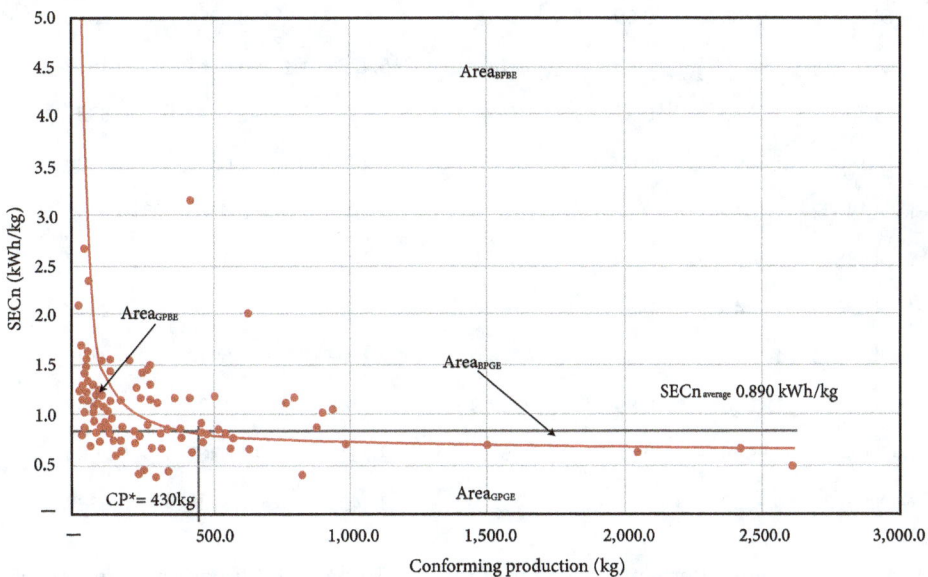

Figure 4.31 PCCD analysis of the SEC$_n$ from PCLD by reference. Points above the curve indicate inefficient references, guiding efforts to improve process scheduling and reduce energy costs

In Figure 4.30, all the molds above the line make the EAC less energy-efficient. In Figure 4.31, the points above this line are all references that have the same effect on the EAC and thus reduce its energy efficiency. However, the actions to be taken with the molds and references depend on where the points are located, as explained in Figure 4.17.

This example shows the molds with good energy performance and poor energy efficiency. They are the ones that comply with the rule presented in Equation 4.24, and Table 4.6 shows them ordered from lowest to highest energy efficiency. These molds could improve their energy efficiency by increasing the size of the batches produced in each mold. According to Figure 4.30, the minimum recommended production per mold assembly (CP*) is 950 kg, which is the production according to which the $SEC_{n_{average}}$ is achieved. The references that meet the same condition are not presented, but from Figure 4.31, it is possible to establish that the minimum batch size per reference should be higher than 430 kg.

Table 4.6 Consolidated Information on Molds with Good Energy Performance and Low Energy Efficiency

Mold	SEC_n [kWh/kg]	SEC_n average [kWh/kg]	Compliant production [kg]	SEC_n PCCD [kWh/kg]	SEC_n – SEC_n PCCD [kWh/kg]	SEC_n – SEC_n average [kWh/kg]
MOL13739	2.6730	0.89	34.0	6.5028	–3.8298	1.7830
MOL8081	1.3419	0.89	41.6	5.4286	–4.0867	0.4519
MOL706	1.2637	0.89	75.4	3.2831	–2.0194	0.3737
MOL14087	1.1808	0,89	330.9	1.2386	–0.0578	0.2908
MOL14082	1.1645	0.89	365.9	1.1809	–0.0165	0.2745
MOL6883	1.1477	0.89	272.9	1.3668	–0.2191	0.2577
MOL8968	1.1449	0.89	163.6	1.8558	–0.7110	0.2549
MOL14319	1.1279	0.89	126.0	2.2193	–1.0914	0.2379
MOL14037	1.1208	0.89	293.9	1.3145	–0.1937	0.2308
MOL6446	1.0949	0.89	95.3	2.7311	–1.6362	0.2049
MOL8322	0.9289	0.89	71.1	3.4430	–2.5141	0.0389

The most complex intervention is that of the molds that exhibit poor energy performance and low energy efficiency. These molds are presented in Table 4.7, ranked from lowest to highest energy efficiency. These molds require major engineering decisions to improve their EAC performance or move them to other machines where they can perform better.

Table 4.7 Consolidated Information on Molds with Poor Energy Performance and Low Energy Efficiency

Mold	SEC_n [kWh/kg]	SEC_n average [kWh/kg]	Compliant production [kg]	SEC_n PCCD [kWh/kg]	$SEC_n - SEC_n$ PCCD [kWh/kg]	$SEC_n - SEC_n$ average [kWh/kg]
MOL8481	21.2797	0.89	26.7	8.1119	13.1678	20.3897
MOL14575	3.1616	0.89	419.0	1.1117	2.0499	2.2716
MOL7861	2.0254	0.89	632.4	0.9510	1.0744	1.1354
MOL4341	1.5482	0.89	236.7	1.4788	0.0694	0.6582
MOL15202	1.4817	0.89	945.3	0.8465	0.6352	0.5917
MOL15282	1.4027	0.89	715.2	0.9144	0.4883	0.5127
MOL15203	1.2290	0.89	955.5	0.8442	0.3847	0.3390
MOL15229	1.1820	0.89	510.2	1.0265	0.1555	0.2920
MOL15216	1.1718	0.89	801.0	0.8845	0.2873	0.2818
MOL6864	1.1504	0.89	604.3	0.9656	0.1847	0.2604
MOL15207	1.1153	0.89	771.5	0.8941	0.2212	0.2253
MOL8964	1.0432	0.89	938.6	0.8480	0.1952	0.1532
MOL3979	0.8915	0.89	1,929.8	0.7387	0.1528	0.0015

In other words, the increase in energy efficiency can be as simple as increasing the number of compliant products produced by the EAC for the analyzed criterion. In this sense, it is mainly the responsibility of the commercial or production scheduling area. The improvement of energy performance requires engineering, and this is an achievement usually attributable to the plant that manages the EAC. However, adequate engineering practices can help reduce batch sizes to keep the energy efficiency of the EAC under control. As shown in Figure 4.32, if the PCLD could meet the values of the ABTD, batch sizes per mold assembly can drop from 950 kg to about 300 kg, making the plant more flexible to meet commercial challenges.

Additionally, PCLD and PCCD were determined for SEC_s. Since there is no record of the EAC mass flow for each mold during steady-state operation, the steady-state demand was correlated with the effective productivity obtained from Equation 4.32. The result is presented in Figure 4.33, which also presents an acceptable correlation ($R^2 = 0.77$).

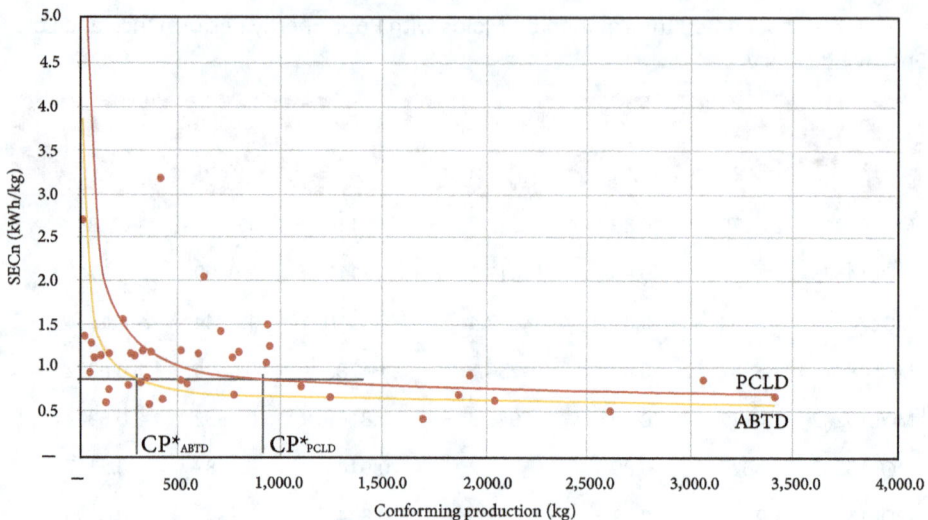

Figure 4.32 PCCD analysis of the SEC$_n$ from PCLD by molds indicating the potential for improvement if ABTD is achieved. This diagram supports strategic planning for mold utilization and process flexibility

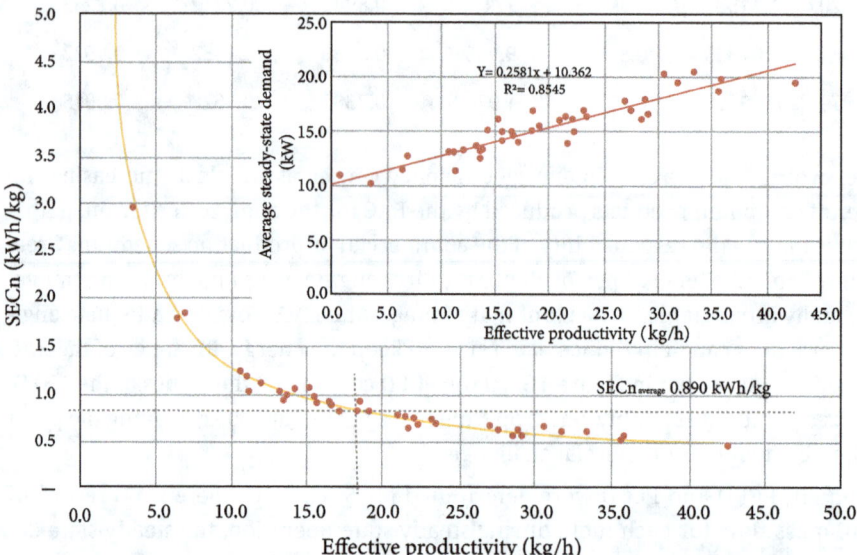

Figure 4.33 Diagram of SEC$_s$ under steady-state conditions plotted vs. effective productivity per mold. The chart reveals the minimum productivity threshold required to maintain acceptable energy efficiency levels

From Figure 4.33, the variable consumption is 0.2851 kWh/kg, and the fixed demand is 10.362 kW. But more importantly, to maintain a consumption lower than 0.890 kWh/kg, the molds must have an effective productivity higher than 18 kg/h.

In this context, the following conclusion is beneficial for process analysis. Suppose both the energy consumption of the injection molding machine and the cost of energy consumption of the products produced will be controlled. In that case, production scheduling should concentrate on three factors:

1. When a mold is assembled, it should be done for production runs above 950 kg

2. When several references are assembled in the same mold, the production batch size per reference must exceed 430 kg

3. Molds that cannot run at an effective productivity of more than 18 kg/h should not be mounted.

These results respond to the specific behavior of the composition of the products being assembled on the machine and the characteristics of the machine. It is worth emphasizing that the cost of energy is not the only factor to consider, but the example shows how to consider the cost of energy in the decision. Finally, the analysis allows modeling the cost of energy invested in the product according to the size of the batch assembled per mold and reference, according to the level of productivity in the process. Based on some equations in previous chapters and Equation 4.30, the following product costs are obtained:

Product cost per SEC_n per mold:

$$\text{ECP} \left[\frac{\$}{\text{kg}} \right] = 0.6353 \left[\frac{\text{kWh}}{\text{kg}} \right] + \frac{199.63 \, [\text{kWh}]}{\text{Production}_{\text{compliant}} \, [\text{kg}]} \times \text{EC} \left[\frac{\$}{\text{kWh}} \right] \qquad (4.35)$$

Product cost per SEC_n per reference:

$$\text{ECP} \left[\frac{\$}{\text{kg}} \right] = 0.6359 \left[\frac{\text{kWh}}{\text{kg}} \right] + \frac{81.54 \, [\text{kWh}]}{\text{Production}_{\text{compliant}} \, [\text{kg}]} \times \text{EC} \left[\frac{\$}{\text{kWh}} \right] \qquad (4.36)$$

Product cost per SEC_s per mold:

$$\text{ECP} \left[\frac{\$}{\text{kg}} \right] = 0.2581 \left[\frac{\text{kWh}}{\text{kg}} \right] + \frac{10.36 \, [\text{kWh}]}{\text{Production}_{\text{compliant}} \, [\text{kg}]} \times \text{EC} \left[\frac{\$}{\text{kWh}} \right] \qquad (4.37)$$

At a minimum, the product cost will be the highest cost calculated with Equation 4.35, Equation 4.36, or Equation 4.37. It is important to remember that the SEC_s does not consider production and quality inefficiencies. Therefore, the energy cost associated with the production of the product will usually be higher than the calculated cost.

4.3 Closing the Production Energy Gap

To close the EAC production energy gap, uptime must be optimized, and energy consumption during non-productive times must be minimized. In production plants, there are many reasons why productive time is lost, and why extra energy is unavoidably consumed during non-productive times.

4.3.1 Recommendations for Optimizing Production Times

Plastic transformation processes are not very friendly to machine stops. Losses of time and materials during startup and shutdown processes are costly; therefore, startups and shutdowns should be minimized as much as possible. There are usually two types of downtime: unscheduled downtime and scheduled downtime. In both cases, there are many opportunities for improvement.

4.3.1.1 Unscheduled Downtime

Unscheduled shutdowns should be (and usually are) appropriately documented as part of the floor control within industrial plants. Tools such as Pareto diagrams are traditionally helpful in prioritizing the causes that generate unscheduled downtime. Among the most frequent causes are the following:

Equipment malfunction

When equipment stops frequently due to machine malfunctions, it is necessary to question the maintenance management of production equipment, the decisions made regarding the technologies acquired, or the practices of plant personnel that may affect the processing equipment.

- From the theory of maintenance management, industrial plants must set up three types of **maintenance processes** [8]:
  - Corrective maintenance is reactive and deals with damage to equipment as it occurs. These events will always happen, but the time spent on this type of maintenance should be minimized, as it significantly affects downtime and, therefore, the production energy gap. A proper preventative and predictive maintenance program directly reduces the time spent on corrective maintenance
  - Preventative maintenance schedules maintenance tasks before equipment breakdowns occur and do so during production downtime
  - Predictive maintenance implements an adequate monitoring program of production equipment to detect faults or trends in equipment behavior that may lead to failures before they occur.

Neither preventative nor predictive maintenance affect the production energy gap. Energy monitoring of process equipment can provide important information not only for process energy performance but also for predictive maintenance programs.

- **The timing of decisions on the acquisition of production technologies** is critical. With proper electrical and control system assembly, processing equipment must be constructed of wear-resistant steel and properly instrumented to prevent damage. In terms of performance, they must be robust enough to withstand the forces, internal pressures, thermal load, and vibrations during operation, with an adequate balance of capacities, using energy-efficient motors, hydraulic systems, and heating and cooling systems. In addition, it is desirable to have a sufficient supply of spare parts and adequate technical support. All this contributes to the reliability of processing equipment and often makes it more expensive. However, in production processes, fixed costs usually have a much smaller impact than variable costs on the total cost of products [9]. The purchase value of processing equipment falls into the category of fixed costs, while energy gaps are part of the variable production costs, so preventing gaps is a priority. The more that production is oriented towards commodities, the more critical it is to make the right investments, since efficiency and productivity through technology must be considered competitive in these markets.

- **Poor practices of plant personnel** can significantly affect processing equipment, leading to faster failures and unscheduled shutdowns. Some of them are:
  - Cold starts: Cold starts occur when the screw of the plasticizing units is driven without the material inside having reached the recommended melt temperature or when the heating zones of the screw and die have not reached the temperature established in the profile. This can lead to accelerated wear of the screw and barrel, screw fracture, breakage of gears and pulleys in the gearbox, damage to static mixers and melt pumps, and generation of excessively high pressures that elongate or break adjustment, and assembly screws, among others.
  - Inadequate adjustment of the temperature profile: Properly adjusting the temperature profile requires knowledge of the screw geometry, the polymer transition temperatures, the presence or not of other components in the blend formulation, the required screw rotation speed, the type of plasticizing unit technology, the kind of product to be manufactured, and the distribution of the heating bands along the cylinder. It is often said that there are two types of people who do not know about polymer processing: *those who ask what the process temperature profile should be without knowing this information, and those who answer*. Inadequate adjustment of the temperature profile can lead to a cold start, lead to operation at pressure levels too high for the mechanical strength of the equipment, and generate accelerated wear of the screw or cylinder in specific regions, among others.

■ Going from zero to process speed too fast: The increase in process speed should be gradual. Properly controlled polymer processing equipment typically has protections against excessive pressures, amperages, and/or torques, and protections for the accelerated increase of pressure, amperage, and/or torque. This prevents the equipment from reaching pressures, amperages, and/or torques that could cause damage. If the processing equipment does not have these protections, special care must be taken to increase the process speed. This is so that the operating personnel have enough time to act when a problem arises that could affect the integrity of the machines.

■ The implementation of provisional solutions that become definitive solutions: This is a common practice to solve mechanical, electrical, sensor, and control problems quickly in the plant. However, they usually do not correspond to the recommended solution. Although these solutions may eventually solve the specific issue, they generate the risk of becoming a definitive solution, and over time, the processing equipment becomes disfigured and loses its reliability.

■ Use of inadequate tools and procedures for disassembly, cleaning, and assembly of equipment: Cleaning should be done with soft metal tools such as copper, bronze, or aluminum to avoid scratching the surfaces of dies, molds, and screws. The assembly of parts and components should be done with torque wrenches. All parts in contact should be lubricated with high-temperature grease. Disassembly is recommended on a table with a temperature-resistant rubber surface to avoid placing the parts on the floor. The assembly should ensure that no leaks occur. If they happen, they should be corrected immediately, because high-pressure melts are very abrasive, causing leaks to worsen with operation. The assembly and disassembly instructions provided by the manufacturer must be followed. All this is to prevent failures and to avoid accelerating the wear of the machine components.

Inadequate changeover procedures

The lost time associated with reference changes due to mounting and preparation procedures is usually addressed by minimizing changeovers by increasing batch sizes. However, there is a delicate balance for batch sizes, as they must be large enough but maintain adequate production flexibility and reduce inventory levels. Therefore, if a flexible production scheme is to be maintained, with low turnaround times and reduced levels of product in transit, it is necessary to minimize the time spent on preparation and set-up before and during each assembly.

One of the most widely used and successful methodologies to achieve this is the SMED (single-minute exchange die) methodology [10]. This methodology was initially applied in Toyota assembly plants and is widely used for Formula 1 pit stop operations. SMED consists of four stages:

1. Classification of the operations to be performed into internal operations, which are performed when the machine is stopped, and external operations, which are performed while the machine is moving

2. Conversion of as many internal operations as possible into external operations, either because they are misclassified and can be carried out while the machine is in motion, or because alternatives can be developed that allow this to be done without sacrificing the safety of personnel and equipment. An example is preheating parts for assembly rather than heating them after assembly

3. Organization of external operations to minimize the time required to perform them. This stage may require investment in specialized tools or the design of specific procedures and tools to optimize the execution of the associated tasks. At this point, other supporting methodologies may be helpful, such as the "5S" methodologies (*seiri* = classification, *seiton* = order, *seiso* = cleanliness, *seiketsu* = standardization, and *shitsuke* = discipline), widely used to organize and improve workspaces seeking greater productivity and efficiency [11]

4. Reduction of time spent on internal operations, which usually involves standardizing to reduce parameterization and adjustment times.

Human errors

Human errors are identifiable actions that cause undesirable results and manifest as problems [12]. Human errors are one of the leading causes of product quality problems and downtime. Implementing a methodological approach to minimize human errors and identify them before they generate a problem is vital because human errors are inevitable, but their effects are not.

There are usually two types of errors: random errors and systematic errors. Information, training, education, and personnel training must be reinforced when errors are random. When errors are systematic, it is usually because some aspect of the work situation causes them, and the focus is on identifying the causes to eliminate the source of the error [13]. Several methodologies have proven successful in manufacturing in minimizing or removing systematic errors. Among them are process standardization, automation of repetitive processes, probabilistic risk analysis, feedback, continuous improvement culture, identification and elimination of root causes, and design of operations and workstations to minimize the probability of error occurrence.

One of the principal methodologies is KAIZEN, a business management strategy that seeks continuous improvement of the organization's operations through daily and progressive improvement to reduce inefficiencies and build an effective and productive environment that increases the company's competitiveness [14]. Regarding identifying and eliminating root causes, one of the most widely used methods is the SHERPA method (systematic human error reduction and prediction approach), which allows the structured identification of error modes associated with specific operations in the execution of a task [15].

Other methods that can be used to identify the causes of error are HFACS (human factors analysis and classification system) and PHECA (potential human error cause analysis) [16]. The 5S approach is instrumental in the design of operations and workstations to minimize the probability of error occurrence. The proper implementation of 5S generates an environment that provides operators with the necessary tools to complete their tasks. It eliminates activities and tools that do not add value or may induce error, and creates a visual workspace where any deviation is easily recognizable [17].

Another widely used technique is the POKA-YOKE method. The translation means "error-proof" or "avoidance of inadvertent errors." Four types of POKA-YOKE can be used independently or combined according to the company's needs:

- Sequential POKA-YOKE implements systems that preserve order and do not allow omissions

- Informative POKA-YOKE develops mechanisms that provide data to prevent errors

- Grouped POKA-YOKE develops kits so that no element that prevents the correct operation of a process is forgotten

- Physical POKA-YOKE develops devices or mechanisms to identify physical inconsistencies in operations or products to avoid errors [18].

4.3.1.2 Scheduled Downtime

The reference changes are the primary source of scheduled downtime affecting the production energy gap. These reference changes should be minimized. As previously discussed, there must be a balance between flexibility and the cost of in-process inventories. The "just in time" approach has the advantage of minimizing in-process inventories but maximizing changeovers. Finding the right break-even point can be complex when budgeting costs, depending on factors such as products and equipment used for processing. From an energy consumption point of view, plastic converting processes benefit from long runs. Some common questions in production plants are:

- How small can the production batch be without significantly affecting the cost of the product due to the energy costs invested in the transformation?

- What is the cost of energy used in the production of each product?

Energy cost is not the only criterion for defining lot sizes; it is also an essential factor. Energy monitoring of production equipment can be handy when cross-checking consumption data with production data to answer the questions posed. The EGM can provide a way to establish both the recommended production batch size and the associated energy cost, even allowing the cost of energy used in production to be individualized and passed on to each product. This is shown in detail in Section 4.2.

4.3.2 Recommendations for Reducing Energy Consumption during Non-Productive Times

The priority will always be to reduce downtime. Downtime can be significantly reduced, but it cannot be avoided. Additionally, startup and shutdown procedures are events during which non-productive energy is consumed. The challenge is to reduce energy consumption during these times. This is the reason for the following recommendations, which should be carried out even if the shutdown is not expected to be prolonged (since events beyond the plant's control can lengthen many shutdowns):

Turn off and do not operate the machine's moving parts until the startup

It is common to find equipment that is not producing but where the motors in charge of the movements are still energized and driven, in parts such as rollers, clamps, conveyor belts, gates, and stackers. All unnecessary consumption is significant and must be eliminated. This is the first action to be taken when equipment stops production.

Close all valves on systems using compressed air

Compressed air in production plants is expensive, and consumes a significant amount of energy. Even if the equipment is stopped and does not appear to be consuming compressed air, it is advisable to ensure this by closing the compressed air supply valves to the machine and opening them only at restarting.

Close the cooling water valves

Cold water flows to machine cooling systems such as molds, chill rolls, and heat exchangers should be suspended. These devices will continue to remove heat from the environment and generate unnecessary consumption in the chiller and cooling towers. It is estimated that cooling to sub-ambient temperatures can be up to three times more expensive than heating. Water flow should not be suspended in the temperature control systems or the plasticizing units' feed zones, because it causes problems associated with the plasticization of the material in the feed zone or its accelerated degradation in other zones of the screw, which will only lengthen the shutdown.

Establish "stand by" temperature profiles for the plasticizing units of injectors, extruders, and blow-molders

It is recommended to use the "stand by" temperature profile to warm up the equipment before the plant starts production and to adjust it when production stops due to an unscheduled machine shutdown. A proper "stand by" temperature profile requires adjusting heating zone temperatures higher than the melting temperature for semi-crystalline polymers or higher than the glass transition temperature when working with an amorphous polymer. Adjustments should be made to the corresponding transition temperature [19, 20].

This measure is helpful because it limits heat losses to the environment and thus the energy consumption to maintain the cylinder temperature, since heat losses to the environment are more significant as the surface temperatures of the barrel and die increase. On the other hand, adjusting to the transition temperature reduces the time required to stabilize the process during machine startup since it considerably reduces polymer degradation.

Care must also be taken to avoid a cold start. This requires sufficient time to return to the operating temperature profile to ensure the polymer reaches the desired melt temperature. Plasticizing units can get the desired temperature relatively quickly (20–40 minutes, depending on the size of the machine). Time is more variable with the dies, since simple and small heads take 20–30 minutes, but large and complex heads, such as multi-layer coextrusion heads, can take considerably longer. For this, it is essential to ask for the equipment manufacturer's recommendations.

Develop an adequate procedure to shut down processing equipment

Sometimes, processing equipment must be shut down for long periods, either because of extended corrective maintenance or because the machine has no production schedule. A proper shutdown procedure for processing equipment seeks to minimize the time for startup and subsequent stabilization of production. The cost–benefit ratio of the recommendations should be evaluated for each case. Some of them are:

- Use a more thermally stabilized polymer charge to stop: This seeks to make the material left inside the machine more resistant to degradation, since it will be subjected to high temperatures and residence times. To achieve this, a master batch of anti-oxidizing agents is recommended. On the other hand, if the plasticizing unit works with polymers with a high tendency to degradation, such as PVC, EVOH, PVDC, and PET, it may be advisable to use a more thermally stable polymer in the last load. Suppose a processing aid such as a fluoropolymer is added to this charge. In that case, the adhesion of the material to the walls will be reduced, and the residence time distribution will be narrowed, facilitating the cleaning of the plasticizing unit during startup. In this way, it is expected that when production restarts, the time required to stabilize the process and obtain the expected product quality will be reduced while reducing the waste or scraps generated.

■ Rapidly lower the temperature of the cylinder: The aim is to reduce the thermal load on the polymer inside the plasticizing unit. To achieve this, it is necessary to set the temperature profile reference of the cylinder to a temperature close to the ambient temperature after the cylinder has been emptied. This will activate the cooling system of the heating zones and quickly reduce the temperature. When the temperature of each heating zone is below the glass transition temperature for amorphous polymers or below the melting temperature for semi-crystalline polymers, the machine can be completely shut down to avoid unnecessary energy consumption.

■ Empty the plasticizing unit: Leaving the plasticizing unit full makes it much more time-consuming to reheat. In addition, it makes it difficult to remove the residual and degraded mass that remains in the plasticizing unit when it stops.

Heat equipment at the right time and to the right temperature

Many polymer processing plants shut down on weekends and start at the beginning of the following week. There is usually a person in charge of starting up the equipment hours before the first shift of the week. Many mistakes are often made when turning on the equipment. The first one is to turn on the heating of the plasticizing units using the work or operation profile. Using the "stand by" temperature profile previously described is advisable. The second is to start heating too early. In this sense, it is essential to properly study the equipment since not all its components require the same time to reach the "stand by" temperature profile. Complex dies likely need more time than the plasticizing units, so the heating bands of these dies are turned on before those of the plasticizing units. In addition, large units are turned on first and smaller ones later.

It is, therefore, advisable to generate a clear protocol for the startup sequence of the equipment, establishing the appropriate order and time to carry out the operation. This also applies to the startup of equipment that has been turned off for other reasons.

4.4 Closing the Quality Energy Gap

The quality energy gap is one of the most difficult to close, since the causes that generate nonconformities can have several origins. One of them is from the very meaning of quality. From the technical point of view, quality is a contract with the customer expressed in product properties and specifications, whose values and acceptable variations must be met. From this point of view and in practice, companies do not sell products; they sell properties and specifications. Considering quality as a contract implies negotiation. The product will not be sold if the contract is not fulfilled and quality is not guaranteed. It is, therefore, important that all product properties and specifications are explicit and documented. It is more common than many companies are willing to admit that many customer returns are due to non-compliance with a property or specification that was not made explicit in the negotiation. The quality area of a company is not created to establish the properties and specifications that the product must meet. It was designed to develop strategies to validate the fulfillment of the contract with the customer. First, speaking of "the quality of the company's products" is inaccurate as an absolute concept. If this quality does not meet customer requirements, the product will not sell. On the other hand, if the quality significantly exceeds expectations, it usually implies additional costs associated with extra controls, trimming, and reprocessing, which increase the quality energy gap and impact both the competitiveness and profitability of the company.

Even when the company manages the negotiation process appropriately, various factors may still lead to contract non-compliance and, consequently, to an increase in the quality energy gap. These causes can be associated with human error, deficiencies in product design, inadequate processing conditions, equipment malfunction, or technological limitations in meeting the established quality standards. Regardless of the reason behind quality issues, the plant needs to develop internal procedures and apply effective methodologies for the early detection of defects and timely correction.

Quality management in manufacturing is approached from two perspectives: reactive and proactive.

- Proactive quality management has a preventative approach, anticipating problems and risks and performing a continuous assessment for decision-making at all process stages to evaluate and contain situations before they become problems

- Reactive quality management has a corrective approach; it reacts to incidents, issues, and risks as they arise. Reactive management evaluates existing problems and directs efforts to the most problematic areas, addressing difficulties and challenges after they have arisen.

An adequate quality system must develop methodologies and tools to leverage both approaches. Adequate proactive management reduces the time spent on solving and the impact of quality problems during manufacturing processes and thus helps to minimize reactive management. The balance between the two will depend on the stage of the product cycle, as shown in Figure 4.34.

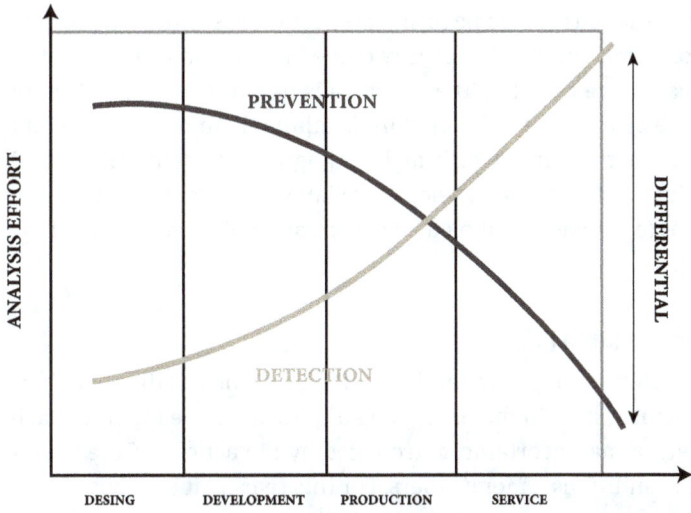

ANALYSIS EFFORT

PREVENTION

DIFFERENTIAL

DETECTION

DESING   DEVELOPMENT   PRODUCTION   SERVICE

STAGE OF THE PRODUCT CYCLE

Figure 4.34 A representation of the allocation of effort in proactive, preventative, and reactive quality control strategies throughout the product lifecycle. The diagram emphasizes the importance of early-stage interventions – particularly during design and development phases – to reduce the occurrence and impact of quality issues during manufacturing, thereby minimizing energy waste and production inefficiencies linked to non-compliant products

4.4.1 Tools for Proactive Quality Management

Some valuable tools for proactive quality management are as follows.

Failure mode and effect analysis (FMEA)

The FMEA method is a technique that allows one to identify, prioritize, and evaluate the causes and effects of failures during the design stages of products and processes, while allowing the establishment of preventative actions to avoid failures and make products and processes more competitive and safer. It is a tool used in "lean manufacturing" to diagnose and prevent errors. The prioritization of actions on the detected causes is based on the evaluation of the risk priority number (RPN), for which the following must be established:

- The probability of occurrence, or frequency with which the failure may occur due to that cause

- The magnitude of the effects or severity of the problem generated by the failure

- The difficulty to detect it.

Preventative or corrective actions should be formulated for causes with a high RPN to reduce the frequency of occurrence or the severity of the failure, and facilitate detection, since these actions reduce the RPN. The objective is to minimize the RPN until there is no risk of failure or until the risk of failure is minimal. Among the inherent advantages of the FMEA method, it is worth highlighting that it promotes detailed knowledge of the product and process, develops a reliable measurement system to detect failures, and helps to formalize complete documentation of the evolution of the product or process [21].

Statistical process control (SPC) [22]

An industrial process is always subject to random variations, such as the variations of the temperature control system in the processing of polymers, the inherent variations of the raw material, the uncertainty introduced by the actions of the operator, and environmental conditions, among others. For this reason, it is impossible to produce two products exactly alike. Therefore, product characteristics and properties are not uniform and exhibit variability, but an adequately controlled process keeps these variations within acceptable limits for the product. This is the objective of SPC.

The statistical basis of SPC is based on the central limit theorem. This theorem states that if a random variable (such as the thickness of a film, the weight of a part, or other property or characteristic of interest) is obtained as a sum of many independent causes, each being of minor importance concerning the whole, then its distribution is asymptotically normal. This means it must have a mean value and a variance from the mean following a normal distribution. All these causes that allow the fulfillment of the central limit theory are called common causes. If the random variable is only influenced by common causes, it is possible to predict the range of variations of the random variable. Everything that occurs within this range of variation due to common causes is considered a chronic problem.

In contrast, if a cause with a preponderant effect on the random variable eventually occurs, this will no longer have the behavior of a normal distribution (for example, those caused by damage to the machines) or will generate variations on the behavior of the normal distribution in which the random variable will have another average and other levels of variation (for example, due to a change of batch of raw material with different properties). These causes are called assignable causes. Variations generated by assignable causes are called sporadic problems. A process is said to be under statistical control when no assignable causes are present. SPC provides statistical tools to analyze the information provided to detect the presence of assignable causes using control charts, such as the one shown in Figure 4.35 [23].

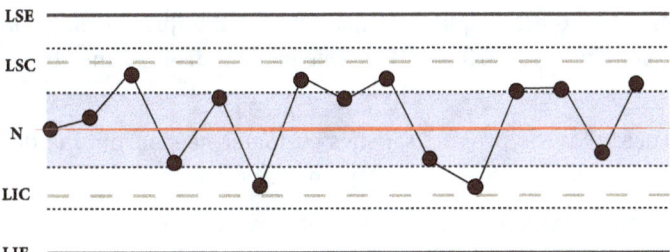

SPECIFICATION AND CONTROL LIMITS

Figure 4.35 Typical control chart used in SPC, illustrating random variable values over time. The graph includes control limits (LIC and LSC) derived from process variability and specification limits (LSL and USL) negotiated with the customer. The chart helps detect assignable causes and assess process capability, including six-sigma compliance

Figure 4.35 plots the value of the random variable on the y-axis, while the x-axis is a time series. LIC and LSC are the lower and upper control limits, respectively, and are statistically determined from the behavior of the normal distribution of the random variable. LSL and USL are the lower and upper specification limits of the random variable, and result from the negotiation process with the client about the value and allowable variation of the specification or property. When the control limits are within the specification limits by a wide margin, the process can ensure production within the specification limits. If $\bar{x}$ is the mean of the random variable (denoted N in Figure 4.35) and σ is the standard deviation, when the specification limits are USL $= \bar{x} + 3\sigma$ and LSL $= \bar{x} - 3\sigma$, the probability of the product going out of specification is 3.4 in a million. This is achieved by negotiating with the customer this variation or improving the process until the standard deviation allows the condition to be met. This variation is called the six-sigma variation.

Therefore, SPC makes it possible to measure the capability of the process to deliver a specification reliably and does so through the calculation of an index called C_{pk}. A capable process must have $C_{pk} > 1$. To achieve six-sigma variation, C_{pk} must be > 1.5 [23].

On the other hand, another utility of control charts is their ability to prevent the production of non-compliant products by monitoring the behavior of the random variable. In a process under statistical control, the behavior of the random variable should be random around the mean. When the time series shows trends, such as more than four consecutive data increases or decreases, the variable is no longer random, and its variations are due to some assignable cause that must be found and controlled. For this preventative function to be practical, control of the random variable must be done at the work center and in line with production.

Juran trilogy

The methodology that best uses proactive quality management is the "Juran trilogy for quality management" or simply "Juran trilogy" [24]. This method distinguishes three interrelated processes:

- Quality planning ensures that the product satisfies customer needs from the design stage, where the FMEA methodology can play a decisive role

- Quality control is based on SPC to detect early anomalous deviations of the random variable with a tendency to go out of specification limits or that directly does not comply with them. In addition, quality control seeks to provide feedback to those responsible for the process to undertake the necessary corrective or preventative actions to eliminate the assignable causes and return the process to a state of control within the conformity zone

- Quality improvement involves initiatives to improve product quality and process performance by reducing the chronic variability of the random variable. Typically, the latter requires investments that must be made at a competitive cost [24], as shown in Figure 4.36.

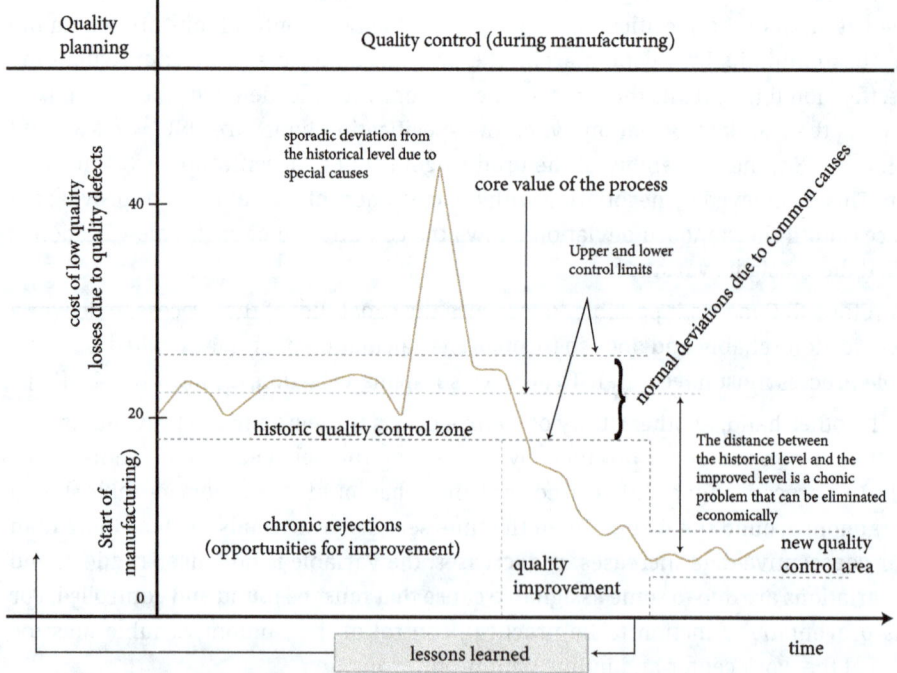

Figure 4.36 The Juran trilogy approach to addressing sporadic and chronic quality problems through three interrelated stages: quality planning, control, and improvement. The figure distinguishes how each component acts on different sources of variation to maintain or enhance product quality and process consistency [24]

4.4.2 Tools for Reactive Quality Management

Some valuable tools for reactive quality management are the following:

Inspection of batches by sampling

Quality inspection is a systematic process responsible for establishing the conformity of products with established acceptance criteria by determining one or more product characteristics using standardized procedures known as quality standards. The acceptance criteria are part of the quality contract with the internal or external customer. In contrast, the quality standards are the standardized testing tests through standards that are international (DIN, ASTM, and ISO, among others), national, the company's, or the customer's own [25].

The inspection by sampling inspects the quality of a random sample of a specific size using different criteria, so inspecting 100% of the products is unnecessary. The larger the sample size, the higher the detection costs, but the lower the costs associated with the defective products shipped, as shown in Figure 4.37. Two typical errors can occur since it is a sample, not the total batch population:

- Type I errors are made when rejecting a product that complies with the specifications

- Type II errors are made when accepting a non-compliant product [26].

The sample size should be such that it has statistical significance to minimize these errors within a certain margin of confidence or probability. The sampling technique should ensure that the sample is random (unbiased). The recommendations of ISO 2859 [27] or the military standard MIL-STD-105E [28] can be used for size determination and acceptance/rejection criteria regarding the number of defects found. Both standards present sampling procedures and reference tables for inspection by attributes.

There are usually three levels of inspection: rigorous (LEVEL I), regular (LEVEL II), and reduced (LEVEL III). Regular inspection is the most used and recommended at the beginning of the inspection process. A rigorous inspection determines compliance with critical quality variables or when the regular inspection shows more defects than expected. Reduced inspection can be implemented for non-critical quality variables or when there is confidence in the quality of the product or process. The referenced standards establish the procedure to move from one level to another as the inspection process demonstrates the reliability of the product quality [27]. Quality inspection is not a procedure for quality improvement or reduction of the quality energy gap, but other tools can support quality improvement based on inspection results.

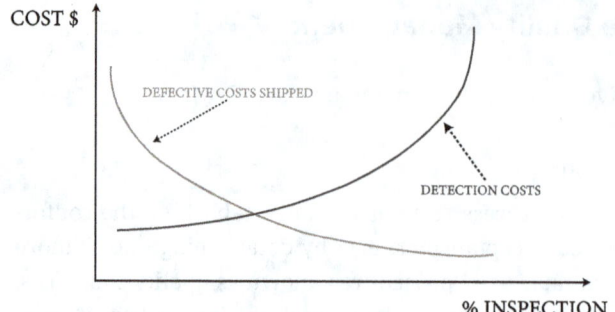

Figure 4.37 Cost analysis of quality inspection strategies, showing the trade-off between detection costs and costs from defective products sent to customers [28]. The figure supports optimal inspection levels based on risk tolerance and product criticality

The eight disciplines (8D) and the problem-solving log

8D is a team-based analysis approach that requires the sequential execution of eight steps, which are as follows:

1. Form a work team: This team should be constituted of people with sufficient knowledge about the process or product to be intervened, and coming from different departments of the company, to offer different points of view and enhance the prospects of finding the best solution

2. Define the problem: This step should provide a detailed description of the problem on which the team's actions will focus. Usually, an adequate description of the problem provides an answer using as much information as possible about what, when, how much, who, how, where, and why

3. Implement containment actions (corrections or interim actions): These actions are intended to mitigate the effects of the problem immediately so that the cost of non-quality does not continue to grow

4. Determine the root cause of the problem: The work team seeks to determine the cause of the problem being addressed

5. Determination of permanent corrective actions: This step involves actions designed to eliminate the root cause definitively. It is the planning step during which tasks, responsible parties, and plan implementation dates are assigned

6. Implement corrective actions: Implementing corrective actions should include not only the actions to solve the problem but also properly document the solution through the necessary changes in processes or procedures, the modification of risk matrices, and the execution of the required training and dissemination plans

7. Implement preventative actions: This step extends the analysis to other similar products or processes where the same problem may occur

8. Congratulate and recognize the work team.

The problem-solving logbook becomes an excellent way to document each step to solve the problem, register the actions taken (with or without success), and allow follow-up, feedback, and continuous improvement. It also becomes an excellent manual of lessons learned for the company.

Cause–effect diagrams

The cause–effect diagram, Ishikawa diagram, or fishbone diagram, is a graphical representation that helps to identify the possible causes of a quality characteristic inconsistency or defect [29], such as the one shown in Figure 4.38. The central spine represents the problem or undesired effect, and the side spine represents the possible causes, which are usually grouped into six categories known as the "6M": Causes associated with man (workforce), method, material, machine, measurement, and mother earth (Environment) [30]. This identification helps to evaluate the causes and to establish action plans to solve the problem.

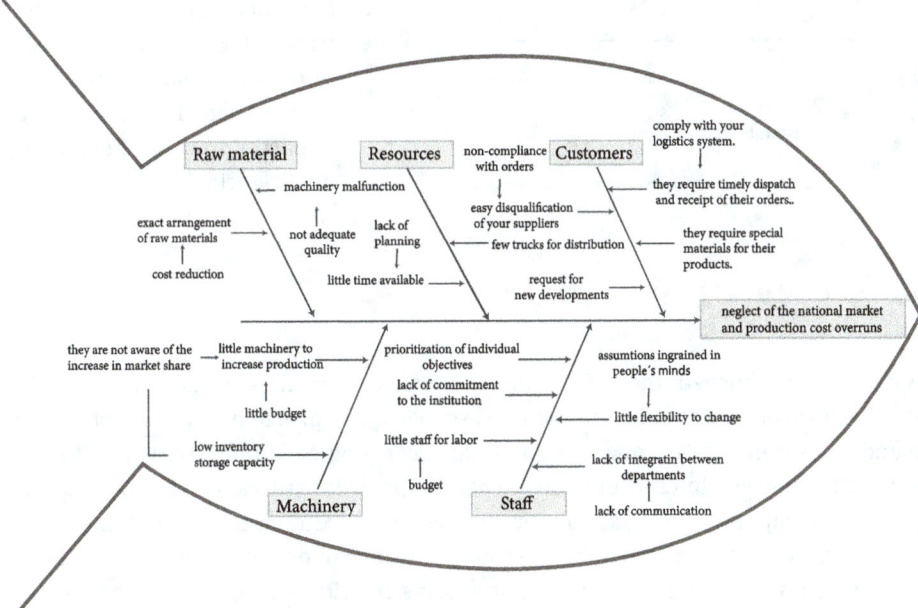

Figure 4.38 A typical Ishikawa or fishbone diagram illustrating a structured approach to identifying root causes of quality problems

Pareto diagrams

The Pareto diagram, also known as the ABC distribution curve, makes it possible to classify the causes of a problem from most to least frequent to determine the leading causes to be solved (see Figure 4.39), on the basis that less than 20% of the causes are usually responsible for more than 80% of the times that a quality problem appears. In this way, action plans can be generated that prioritize eliminating or reducing the leading causes [31].

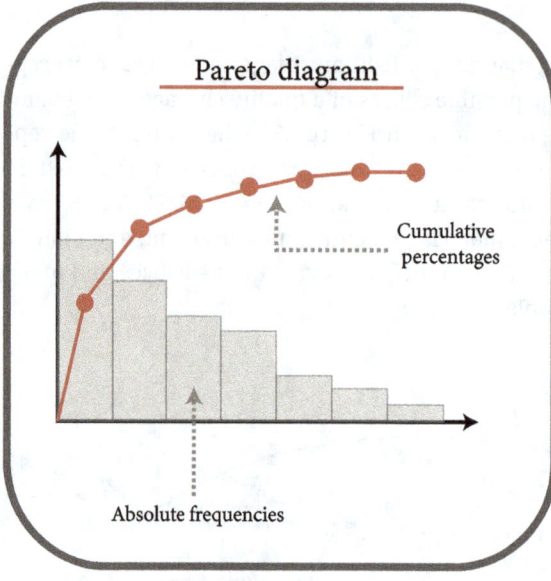

Figure 4.39
Pareto diagram model for quality control, ranking problem causes from most to least frequent. Based on the 80/20 principle, it helps prioritize actions that address the most impactful sources of defects or inefficiencies

4.5 Closing the Process Energy Gap

As already established, the SEC_m is the SEC_s of the machine and the process, under steady-state conditions at maximum process speed. The problem is that the power demand measurements are made at the usual processing conditions, so it is feasible to determine the SEC_s, but it is unknown whether the usual process condition is the best processing condition for maximum productivity or maximum throughput. If it is, the process gap does not exist. If it is not, some strategies can be put in place to estimate it, but by the time it is confirmed which processing condition maximizes process speed, the process gap has also been closed. Typically, when tracking machine and process energy performance, that condition is unknown and will not be known until process optimization methods are applied. This can change if the methodology has been applied to the machine for a long time and its energy performance is already well-known.

The EGM is a method of diagnosing energy inefficiencies to establish action plans to close them. During the diagnostic process, what is usually possible is to determine the combined process + technology energy gap, as shown in Figure 4.40.

When this process + technology gap is significant, it becomes essential to address the process energy gap, determine its magnitude, and decide whether it should be addressed. Processing conditions must be found to maximize the process speed and reduce the process energy gap. When these conditions have been found, the SEC_s is determined and this value becomes the SEC_m of the machine. It is worth clarifying that the SEC_m can be different for different products produced in the same EAC, so the diagnosis and intervention are performed prioritizing the most representative products of the line.

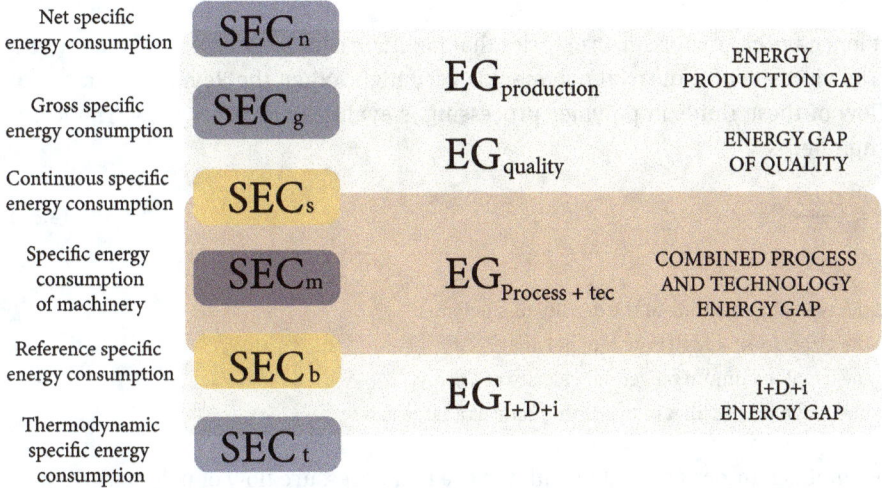

Figure 4.40 Integrated analysis of energy inefficiency using the EGM, representing the combined process and technology energy gap. This diagnostic tool supports strategic decision-making for energy optimization in polymer processing systems

Finding these conditions is a challenge. First, however, it is crucial to understand how energetic transformations occur in polymer processing. For them, looking at what happens inside the control volume is necessary. The flow of polymers in the processes of injection, extrusion, and blow-molding of hollow bodies (which together account for most of the processed polymers in the plastics industry [32]) occurs using a plasticizing unit. The phenomenon depicted in Figure 4.41 is presented by simplifying what occurs in fluid transport in a plasticizing unit and the combined flow of drag and pressure between two parallel plates of a polymer fluid that adheres to both plates.

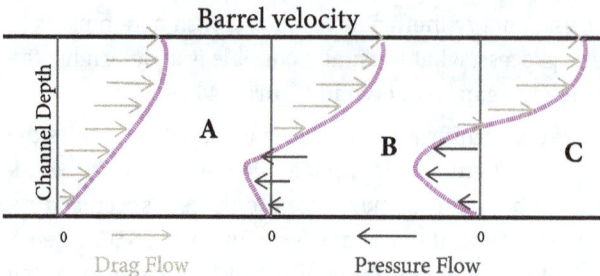

Figure 4.41 Schematic representation of polymer melt flow between parallel plates, combining drag and pressure-driven components. The diagram illustrates laminar flow behavior in plasticizing units, relevant for understanding energy dissipation and homogenization challenges

Polymers have characteristic properties that facilitate some processes and hinder others. The viscosity of industrial polymers is very high. When the Reynolds number of the flow of these fluids in polymer processing is evaluated, it is very low. The Reynolds number is:

$$\text{Re} = \frac{\rho v H}{\eta} \tag{4.38}$$

where:

ρ is the typical magnitude of the density of a polymer

η is the viscosity of a high flow rate polymer

v is the usual maximum tangential velocity of a screw

H is the height of the fillet in the dosing zone of a large screw.

The Reynolds number associated with the drag and pressure flow of polymers such as that shown in Figure 4.41 is typically less than 1 and therefore less than 2,100. The immediate implication is that the flow of molten polymers is always laminar. This is a characteristic of the flow that hinders the melt homogenization processes. Still, at the same time, it is the characteristic that facilitates the co-extrusion and co-injection processes.

This also implies higher energy expenditure is required to obtain a homogeneous melt. Homogeneity in melting composition benefits from turbulent flows and high diffusion of the mixture components. However, the diffusion coefficients of pigments, additives, and other polymers in the plastic melt are usually very low. If the composition is to be homogenized, it must be "forced" to occur.

The high viscosity and low thermal conductivity of polymers have another implication: more heat is generated and absorbed by viscous dissipation (heat generated by friction between the polymer layers, resulting from laminar flow) than heat diffusion, which benefits from the high thermal conductivity. This is appreciated when evaluating the Brinkman number, which is defined as:

$$Br = \frac{\eta v^2}{k(T - T_0)} \tag{4.39}$$

where:

k is a typical level of thermal conductivity of the polymer

T is the maximum temperature of the melt

T_0 is the temperature of the plates in Figure 4.41.

The critical conditions in Figure 4.41 result in $Br = 10$. This indicates that, for this case, viscous dissipation (the numerator in Equation 4.39) is at least 10 times greater in magnitude than heat diffusion, which corresponds to the denominator. The direct consequence is that the increase in polymer enthalpy during processing is mainly due to viscous dissipation. The more viscous the polymer and the faster the process, the greater the influence of friction between the polymer layers on its plasticization and heating in a plasticizing unit.

The plasticizing unit has two main energy inputs: the energy provided by the main motor and the energy supplied by the heating bands. According to the Brinkman number, more energy is harnessed from the main engine, which is converted to enthalpy through viscous dissipation, than from the heating bands. This is evident in Figure 4.42.

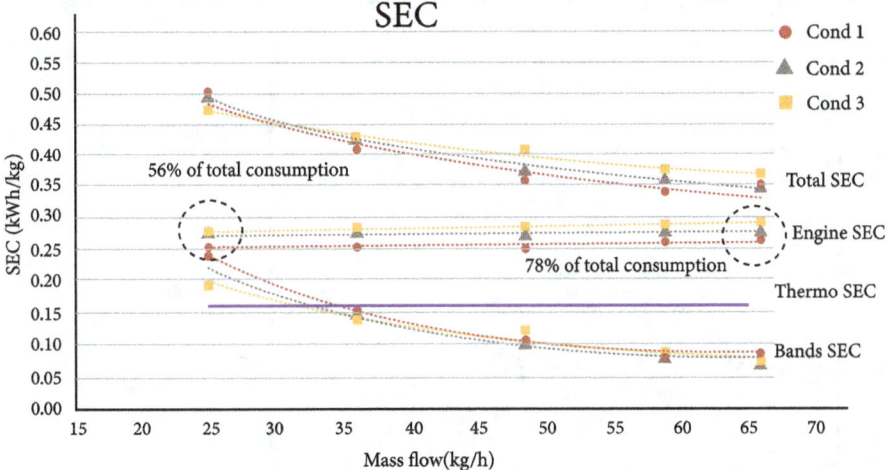

Figure 4.42 Total consumption, motor, and heating band energy consumption as a function of screw rotation speed and screw restriction level, using a 45 mm diameter extruder with grooved feed zone, working with polypropylene. The figure emphasizes the dominance of mechanical energy (from the motor) converted to thermal energy via viscous dissipation

In Figure 4.42, it is possible to observe that the higher the screw rotation speed, the higher the amount of energy consumed by the main motor compared to the energy consumption of the heating bands. This also suggests that, when working close to the maximum capacity of the processing equipment, actions to improve the conversion of the motor's mechanical energy into thermal energy are more important than those that make the energy consumption of the heating bands more efficient.

If the problem of thermal inhomogeneity is not solved, the only way to process the polymer to the quality standards imposed on the product is to reduce the processing speed. If the Brinkman number is analyzed in detail, it is possible to understand it. The heat generated by friction increases with the square of the speed. This means that doubling speed increases the frictional heat by a factor of four. This explains the behavior of Figure 4.42 but also means that it accelerates the occurrence of temperature differences that may become impossible to handle during processing and obtain a stable process. Reducing the processing speed increases the specific energy consumption, as was justified when discussing the monitoring & targeting concepts.

It is, therefore, vital to reduce homogenization problems by "forcing" homogenization to happen. For this reason, the use of barrier screws, with dispersive and distributive mixers and/or static mixers, is a technological standard if you want energy-efficient technology. Obviously, it is not the only factor but one of the most important. Homogenization consumes energy but much less than the energy consumed per kilogram produced when working slowly. Another characteristic of polymers is that the viscosity is pseudoplastic. This means the viscosity decreases with the shear rate increases, as shown in Figure 4.43.

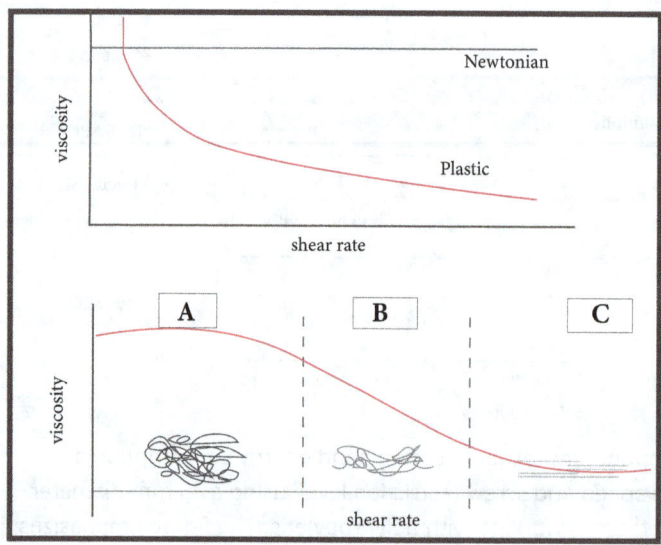

viscosity = shear stress/ strain rate

Figure 4.43 Rheological behavior of thermoplastics, highlighting pseudoplasticity where viscosity decreases with increasing shear rate. The curve supports decisions on machine size and operational speed for energy-efficient processing

In principle, when the viscosity is lower, the energy consumption of the control volume is reduced. To achieve this, one of the actions is to increase the shear rate. Thus, there is a big difference in achieving a target mass flow on a large machine (low screw speed) or a smaller machine (high screw speed). From an energy standpoint, the process will be substantially more efficient in the second case.

The second action would be the increase in temperature. This reduces the pressure required to operate the system. However, the net result is higher energy consumption since, as seen from Chapter 2 (Equation 2.13 and the results shown in Table 2.1), the increase in enthalpy is much more significant than the reduction in pumping power. Therefore, the best decision is always to work at the lowest possible temperature.

In summary, several fundamental conclusions can be drawn from this analysis regarding the operation of polymer processing equipment because of the intrinsic characteristics of these materials:

1. One should operate at the highest possible speed. If two machines can give the same flow, it is advisable to prefer the equipment that operates at a higher rotational speed

2. Homogeneity problems must be solved. This may imply technological changes, but it will allow for an increase in processing speed and, therefore, productivity

3. If plasticizing can increase the temperature profile or rotation speed, it is recommended to favor the rotation speed to reduce energy consumption. This is especially useful in thermoplastic injection molding to take advantage of the pseudoplastic behavior of the polymer

4. The melt should be run as cold as the process allows. It does not matter if the pressure is increased if it is within safe operating limits, since much more energy is consumed in increasing the enthalpy of the melt.

Additionally, many tools can be used to determine the potential for improvement. Some of these tools are bottleneck analysis, operating curve determination, design of experiments, and process simulation.

4.5.1 Bottleneck Analysis

This methodology establishes why productivity is limited to the current value and the percentage of the processing equipment installed capacities being used. Under normal conditions, processes limit their speed when one of the components or capacities reaches a rate close to 100%. When this occurs, this component or capacity becomes the bottleneck. Identifying bottlenecks allows for establishing strategies to remedy the limitation and release the capacity of the equipment to produce faster until a new bottleneck is found. This methodology can be used with any of the polymer transformation processes and requires a detailed study of the process to define the indicators and identify the bottlenecks in each. The method is simple, but it is essential to consider the following recommendations:

1. Make a list of all the components or process variables that can limit the productivity of the polymer processing line

2. Determine the capacity of each component or the maximum allowable value of each process variable on the list. This capacity can usually be determined from the equipment specifications in the operating manual and the parts and machine components supplied by the manufacturer. If this information is not available, it should be measured. The maximum value of the process variables depends on the material's characteristics and properties

3. Establish indicators to evaluate the current use of each component's capacities or the closeness of the process variables to the critical values for each variable in the list

4. Evaluate each of the indicators. The highest valued indicator, or closer to 100%, is considered the bottleneck. Therefore, an action plan should be established to increase that component's capacity or reduce the current value of the limiting productivity process variable.

5. After the improvements are implemented, productivity increase tests should be performed, and the product quality should be evaluated. If the product quality is satisfactory, the indicators should be assessed under the new process conditions to determine the new bottleneck and repeat the procedure from this step.

There will come a point in the improvement process where it will be impossible to continue improving without significant technological changes. From that point on, the remaining energy gap is the technology energy gap. The bottleneck method must be adapted to each process. In the following, the implementation of the technique for thermoplastic extrusion and injection processes will be presented.

4.5.1.1 Determination of Bottlenecks in the Extrusion Process

The typical stages and components of the extrusion process are shown in Figure 4.44.

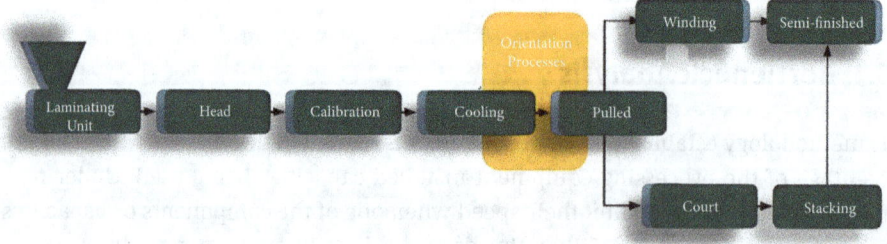

Figure 4.44 Process schematic of a typical thermoplastic extrusion line, identifying the main stages and components. The diagram is used to locate bottlenecks and determine energy and productivity limitations

In the extrusion process, production speed may be limited for one of the following reasons:

▪ The screw has reached the maximum rotation speed

▪ The extruder main motor is limited by demand or current

▪ The process is limited by the mass flow that the plasticizing unit can deliver

▪ The temperature of the polymer (T_m) is either very high or very low in the process

▪ The melt pressure (P_m) is very high in the process

▪ The cooling or heating capacity is insufficient

▪ The available linear speeds of the post-extrusion components are low

▪ There are high variations in mass temperature.

These factors are discussed further below.

The screw has reached the maximum rotation speed

The screw of a plasticizing unit rotates at a maximum rotational speed (N_{max}) that depends on the maximum rotational speed of the motor and the speed reduction ratio of the gearbox and/or pulleys. The use of this capacity is given by the equation:

$$f_N = \frac{N \, [\text{rpm}]}{N_{max} \, [\text{rpm}]} \tag{4.40}$$

The extruder main motor is limited by demand or current

The main motor of a plasticizing unit has a capacity represented in the maximum power (Power$_{max}$) at which it can operate, which is usually stated on the motor nameplate. Equation 4.41 gives the use of this capacity:

$$f_{Power} = \frac{\text{Power} \, [\text{kW}]}{\text{Power}_{max} \, [\text{kW}]} \tag{4.41}$$

where:

Power $= n_{motor} \cdot I \cdot V_{arm}$ for DC motors Power

Power $= \sqrt{3} \cdot I \cdot V \cdot \cos \phi$ for AC motors

I is the current in amperes

n_{motor} is the motor efficiency

V_{arm} is the armature voltage in volts

V is the plate voltage in V

$\cos \phi$ is the power factor.

It is also worthwhile monitoring the current using Equation 4.42:

$$f_I = \frac{I\,[\mathrm{A}]}{I_{max}\,[\mathrm{A}]} \tag{4.42}$$

where:

I_{max} is the maximum current admitted by the motor set on the nameplate

This monitoring is recommended because sometimes the motor works with sufficient power but the limitation is in the current requirements.

The process is limited by the mass flow that the plasticizing unit can deliver

The mass flow delivered by a plasticizing unit depends on the type of plasticizing unit technology, the design of the screws, the polymer used, and the operating conditions.

For single-screw extruders, it is essential to determine the specific mass flow ($\dot{m}_{sp}$). This value can be determined from the characterization of the plasticizing unit by correlating the mass flow rate with the screw rotation speed, as shown in the example schematic in Figure 4.45. In this case, the slope of the straight trend line passing through the origin of the Cartesian diagram is the specific mass flow rate (with a value of 1.9795 kg/(h rpm) in the example). The following equation can evaluate the use of the plasticizing capacity of the extruder:

$$f_{\dot{m}} = \frac{\dot{m}\,\left[\dfrac{\mathrm{kg}}{\mathrm{h}}\right]}{\dot{m}_{sp}\,\left[\dfrac{\mathrm{kg}}{\mathrm{h\cdot rpm}}\right]\cdot N_{max}\,[\mathrm{rpm}]} \tag{4.43}$$

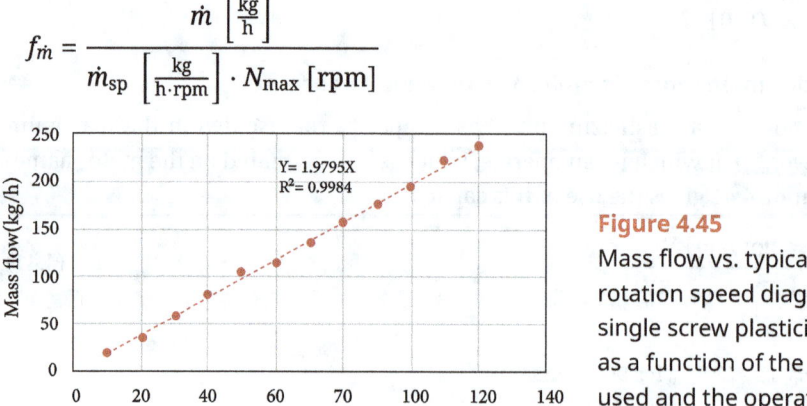

Figure 4.45

Mass flow vs. typical screw rotation speed diagram for a single screw plasticizing unit, as a function of the polymer used and the operating condition

The temperature of the polymer (T_m) is either very high or very low in the process

In the extrusion process, there are two moments where the polymer must be heated: in the plasticizing unit until the material is wholly plasticized, and before the stretching and/or orientation processes presented in the process scheme in Figure 4.44. These stretching processes are only part of some specific extrusion processes, such as the production of bioriented films, mono- and multifilaments, and raffia and biaxially-oriented (bioriented) pipes, among others.

In these cases, the polymer must be heated without plasticizing. However, the proper selection of temperatures for processing is critical. For the definition of these limits, it is important to know the crystallization temperature (T_k) and the crystal melting temperature (T_f) in semi-crystalline materials, and the glass transition temperature (T_g) in amorphous materials. Depending on the temperature, the polymer can enter the thermoelastic or thermoplastic states. Plasticization occurs in the temperature range where the polymer is in the thermoplastic state, and orientation occurs in the temperature range where the polymer is in the thermoelastic state. The thermoelastic range in amorphous polymers begins at a temperature slightly above the glass transition temperature, and ends well before the flow temperature, as shown in Figure 4.46. In this case, the flow temperature is well above this transition temperature.

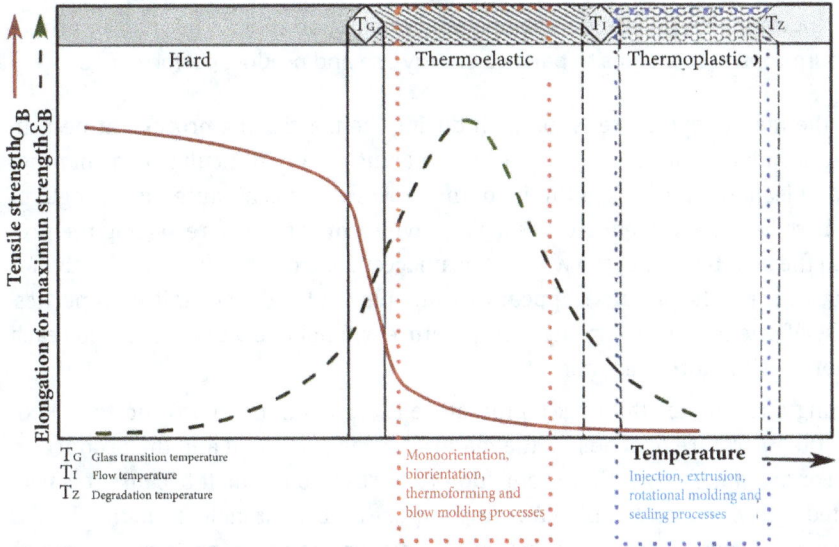

Figure 4.46 Thermal behavior of amorphous polymers as a function of temperature, delineating the glass transition, thermoplastic, and thermoelastic regions. This classification guides temperature selection for energy-efficient plasticizing and orientation processes

On the other hand, as shown in Figure 4.47, the thermoelastic range of the semi-crystalline polymers starts a little earlier and extends to a little after the crystallization temperature. In this case, the thermoplastic state occurs well after the crystal melting temperature.

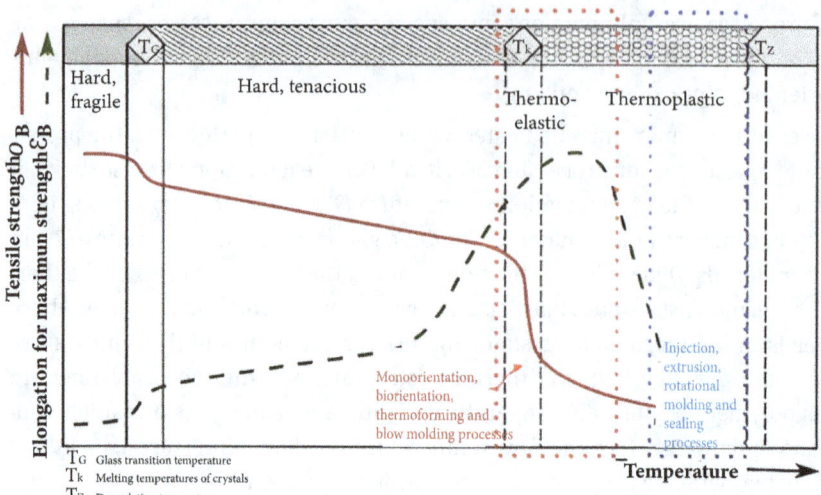

Figure 4.47 Phase state transitions of semi-crystalline polymers depending on temperature, indicating crystallization, melting, and orientation zones. The chart informs temperature control strategies to balance energy use and product performance

Suppose the melt temperature is too high during the plasticizing process, at best. In that case, extrudate control becomes difficult, making it more difficult to control product size and jeopardizing the continuity of the process. This is because the strength of the melt is reduced, and the melt is "sagging", which may require restarting the process when the melt breaks. In the worst case, unacceptable degradation of the polymer can occur, affecting the product's appearance and physical and mechanical properties. In the case of orientation, a very high temperature will not allow the product to reach orientation levels before breaking.

Establishing a ceiling for the bulk temperature ($T_{m,max}$) is imperative, and the maximum melt temperature depends on the material being processed and the production process. For example, blown film production requires a cooler melt because it is usually limited by cooling, and bubble forming requires excellent melt strength. The flat film extrusion process requires substantially higher melt temperatures to ensure glossy films of excellent optical properties or that perfectly copy the cooling roll finish. In the latter process, cooling is considerably faster and more efficient, and additionally, the melt exits the die within a few millimeters of the chill roll. Therefore, low melt resistance is less of a problem. The factor indicating how close the melt temperature is to the upper permissible temperature limit can be evaluated according to the Equation 4.44:

$$f_{T_{m,max}} = \frac{T_m[°C]}{T_{m,max}[°C]} \tag{4.44}$$

On the other hand, when the mass temperature during plasticization is too low, problems of gel generation, the appearance of flow anomalies, homogeneity problems of the mixtures, and an excessive increase of pressure and torque in the system, among others, can occur. Product rupture or cold stretching may happen when the mass temperature is too low during orientation. This generates a plastic deformation that makes it difficult to control the thickness of the product and affects its mechanical properties. For this reason, minimum melt temperature ($T_{m,min}$) must be defined. The factor indicating how close the bulk temperature is to the lower permissible temperature limit can be evaluated according to the following equation:

$$f_{T_{m,min}} = \frac{T_{m,min}\,[°C]}{T_m\,[°C]} \tag{4.45}$$

The approximate values for the limits for the mass temperature come from a rule of thumb that must be used carefully. These heuristic conditions are not always fulfilled because they do not respond to physical laws, but to common qualitative behaviors with exceptions.

The rules of thumb for determining the melt temperature range in the plasticizing unit are:

- For the thermoplastic state for amorphous polymers:

$$T_g\,[°C] + 70\,°C < T_m\,[°C] < T_g\,[°C] + 100\,°C \tag{4.46}$$

- For the thermoplastic state for semi-crystalline polymers:

$$T_f\,[°C] + 40\,°C < T_m\,[°C] < T_f\,[°C] + 70\,°C \tag{4.47}$$

The rules of thumb for determining the range of bulk temperature for mono and biorientation processes is:

- For thermoplastic state for amorphous polymers:

$$T_g\,[°C] + 10\,°C < T_m\,[°C] < T_g\,[°C] + 40\,°C \tag{4.48}$$

- For thermoplastic state for semi-crystalline polymers:

$$T_k\,[°C] - 5\,°C < T_m\,[°C] < T_k\,[°C] + 15\,°C \tag{4.49}$$

Some values of $T_{m,max}$ and $T_{m,min}$ are presented in Table 4.8 for the thermoplastic state and in Table 4.9 for the thermoelastic state. These could be used as a guide depending on the type of polymer and extrusion process.

Processes requiring cooler melts during plasticization may use values close to those presented in Table 4.8 or those calculated from Equation 4.45 and Equation 4.46. These include blown film extrusion, extrusion-blown hollow body extrusion, pipe extrusion, sheet extrusion, and profile extrusion. Processes such as flat-film extrusion, extrusion-coating, and single and multifilament extrusion may require temperatures 20–40 °C higher than in this table.

Table 4.8 Estimated Temperature Range for the Thermoplastic State of Polymers

Polymer	Type	T_g [°C]	T_k [°C]	T_f [°C]	$T_{m,min}$ [°C]	$T_{m,max}$ [°C]
LDPE	Semicrystalline	–120 to –90	85–95	105–115	150	190
HDPE	Semicrystalline	–120 to –90	105–115	120–135	160	210
H-PP	Semicrystalline	–10 to 0	100–120	160–170	190	240
R-PP	Semicrystalline	–40 to –20	80–120	130–160	170	230
I-PP	Semicrystalline	–60 to –50	100–130	130–160	170	230
GPS	Amorphous	80–100	—	—	170	200
HIPS	Amorphous	80–100	—	—	170	200
PET	Semicrystalline	67–81	180–220	240–260	280	310
ABS	Amorphous	~105	—	—	175	205
PC [a]	Amorphous	~150	—	—	280	320
PVC	Amorphous	70–80	—	—	150	180
LDPE	Semicrystalline	–120 to –90	85–95	105–115	150	190

[a] Strictly speaking, PC is a semi-crystalline polymer, but its degree of crystallization is so low that its behavior is more like an amorphous polymer.

In the case of variables with an optimum operating range, the factors are expected not to exceed 100% (value of 1). However, the variables under appropriate process conditions usually give values close to any of the established limits. A very high or very low melt temperature is generally due to an inadequate adjustment of the temperature profile. However, when actions on the temperature profile do not achieve significant effects on the melt temperature, it may be due to an inadequate design of the extrusion screw.

Table 4.9 Estimated Temperature Range for the Thermoelastic State of Polymers

Polymer	Type	T_g [°C]	T_k [°C]	T_f [°C]	$T_{m,min}$ [°C]	$T_{m,max}$ [°C]
LDPE	Semicrystalline	–120 to –90	85–95	105–115	85	105
HDPE	Semicrystalline	–120 to –90	105–115	125–135	105	125
H-PP	Semicrystalline	–10 to 0	100–120	160–170	110	155
R-PP	Semicrystalline	–40 to –20	80–120	130–160	95	150

Polymer	Type	T_g [°C]	T_k [°C]	T_f [°C]	$T_{m,min}$ [°C]	$T_{m,max}$ [°C]
I-PP	Semicrystalline	−60 to −50	100–130	130–160	95	150
GPS	Amorphous	80–100	—	—	110	150
HIPS	Amorphous	80–100	—	—	110	150
PET [a]	Semicrystalline	67–81	180–220	240–260	110	130
ABS	Amorphous	~105	—	—	115	155
PC [b]	Amorphous	~150	—	—	145	190
PVC	Amorphous	70–80	—	—	85	125

[a] The thermoelastic state is required when PET has been processed and crystallization inhibited, which is why it is determined by the transition temperature and cold crystallization (110–130 °C).

The melt pressure (P_m) is very high in the process

An extruder is divided into two components. The first is the plasticizing unit (the screw–cylinder system). The second is the restriction, which includes the die, the screen changer, the screens, the static mixer, and the melt conveyance sections from the plasticizing unit to the die. The first component acts as a pressure builder, while the second is a consumer, as shown in Figure 4.48.

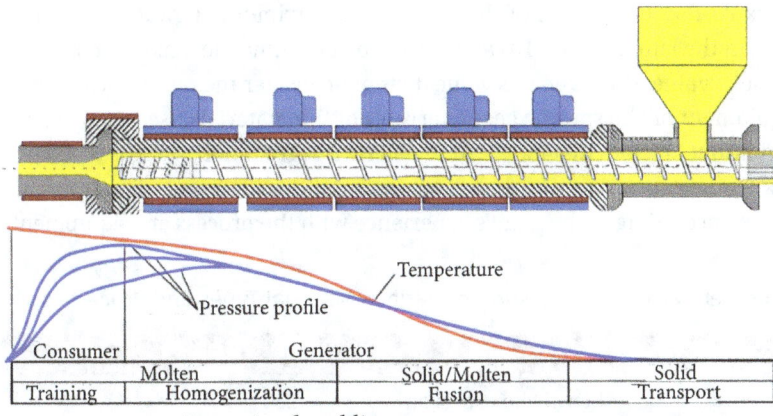

Figure 4.48 Pressure and temperature profile in a thermoplastic extrusion system. The figure illustrates how pressure is built in the plasticizing unit and consumed across the restriction elements, helping to diagnose stability and quality risks

A very low melt pressure usually means that the extruder operates in a suction mode. In other words, this means that stable melt pressure and flow are not achieved, which makes product calibration difficult. In addition, the melt homogenization process is hindered. With this, the mass temperature variation across the flow cross-section also does not help obtain a controlled product in dimensions and physical and mechanical properties. However, low pressures in the system do not limit productivity, since pressure increases with mass flow. On the other hand, low pressure generates quality and process stability problems.

In contrast, very high pressures imply high shear stresses inside the extruder. These can cause flow anomalies or polymer degradation so that the upper limit will depend on the type of polymer being processed. In addition, pressures that are too high endanger the integrity of the processing equipment and the operators. In this case, the maximum pressure level allowed by extruder design must also be considered. The factor indicating how close the melt pressure is to the upper permissible pressure limit can be evaluated according to the following equation:

$$f_{P_{m,max}} = \frac{P_m\,[°C]}{P_{m,max}\,[°C]} \tag{4.50}$$

The $P_{m,max}$ value should be determined carefully, considering the equipment manufacturer's recommendations and the characteristics of the polymer used. For example, even if the manufacturer designed the extruder to work at pressures below 600 bar, it may be difficult to operate at these pressures without the polymer degrading, overheating, or flow anomalies being caused in the process. Additionally, extrusion processes require very high molecular weight materials, such as blown film extrusion. At the same time, others require materials with substantially lower molecular weight, such as mono- and multifilament extrusion. In the case of the former, the maximum allowable pressures will be higher than the latter. Table 4.10 can be used to determine the maximum permissible mass pressure value. However, this value does not consider the mechanical limitations of the equipment or the specific characteristics of the materials used. It also does not consider additional factors that may lead to polymer degradation, such as residence time and mass temperature. Therefore, the maximum allowable mass pressure value should be adjusted according to the plant's experience with the process and equipment.

Table 4.10 Estimated Maximum Pressure for Various Material Types and Processes

Polymer	Process	$P_{m,max}$ [bar]
HDPE	Blown film extrusion	350
LDPE	Blown film extrusion	300
H-PP	Flat film extrusion	250
HDPE	Pipe extrusion	300
PET	Sheet extrusion	250

Polymer	Process	$P_{m,max}$ [bar]
HIPS and ABS	Sheet extrusion	250
PVC	Pipe and profile extrusion	180

Very high pressure may be due to multiple causes, such as a high system restriction because the flow channels through the header are too narrow or too long. High pressure also occurs because the meshes used are of exceptionally high mesh or simply because the capacity of the plasticizing unit is not correctly balanced with the restriction generated by the die. A very low melt temperature and the temperature profile used to achieve it also play a role.

The cooling or heating capacity is insufficient

Cooling is critical to ensure that the entire plasticized mass can be cooled fast enough to obtain a product of the required quality and specifications. This is even more critical in processes such as blown film extrusion, which are inherently limited by cooling. For cooling there are three important variables: the heat to be removed from the product (), the installed cooling capacity ($Q_{cooling,max}$), and the efficiency with which the cooling system makes use of that capacity ($\eta_{cooling}$). The installed cooling capacity usage can be estimated using the following equation:

$$f_{Q_{cooling}} = \frac{Q_{cooling}\,[kW]}{Q_{cooling,max}\,[kW] \cdot \eta_{cooling}} \tag{4.51}$$

Considering that h_{T_m} is the specific enthalpy of the polymer at the temperature of interest and $h_{T_{amb}}$ is the specific enthalpy of the polymer at the environment temperature, we have:

$$Q_{cooling}\,[kW] = \frac{\dot{m}\left[\frac{kg}{h}\right] \cdot (h_{T_m} - h_{T_{amb}})\left[\frac{kJ}{kg}\right]}{3{,}600} \tag{4.52}$$

Information on specific enthalpies as a function of temperature can be found in [19]. These enthalpies can also be measured by differential scanning calorimetry (DSC). Cooling efficiency is very relative to the type of cooling and the specific technology used. Convection air cooling is usually more inefficient than convection cooling in a water bath, and in turn, water bath cooling is more inefficient than water spray cooling. Also, the specific technology used is relevant. In blown film, using a triple lip cooling ring with internal bubble cooling (IBC) is more efficient than a single lip cooling ring without IBC.

The $\eta_{cooling}$ values presented in Table 4.11 are suggested based on the experience of the interventions performed. These values can be adjusted according to the user's process experience.

Table 4.11 Maximum Cooling System Efficiency in Different Extrusion Processes

Process	$\eta_{cooling}$ [%]
Blown film extrusion	75%
Flat film extrusion	85
Pipe and profile extrusion	85
Multifilament extrusion	70
Sheet extrusion	80

Similarly, the use of the heating capacity before the orientation and biorientation processes can be calculated using the following equation:

$$f_{Q_{heating}} = \frac{Q_{heating}\,[\text{kW}]}{Q_{heating,max}\,[\text{kW}] \cdot \eta_{heating}} \tag{4.53}$$

Here, the heat required for heating between T_{m1} and T_{m2} is given by:

$$Q_{heating}\,[\text{kW}] = \frac{\dot{m}\left[\frac{\text{kg}}{\text{h}}\right] \cdot (h_{T_{m2}} - h_{T_{m1}})\left[\frac{\text{kJ}}{\text{kg}}\right]}{3{,}600} \tag{4.54}$$

The available linear speeds of the post-extrusion components are low

The driven rolls of the line have a maximum linear speed ($v_{h,max}$), which is limited by the perimeter and the maximum rotational speed of the roll. The latter depends on the motor speed and the reduction system. The winding system for flexible products or slitting for semi-rigid products also has a maximum speed that defines its capacity ($v_{e,max}$). Other systems, such as corona treatment, marking systems, and stacking systems, among other auxiliaries, also have a maximum linear speed ($v_{aux,max}$) recommended by the manufacturer. The estimation of the use of the installed capacity in these components can be calculated as follows, where the maximum values are established in the respective equipment manuals:

$$f_{v_h} = \frac{v_h}{v_{h,max}} \tag{4.55}$$

$$f_{v_e} = \frac{v_e}{v_{e,max}}$$

$$f_{v_{aux}} = \frac{v_{aux}}{v_{aux,max}}$$

Large variations in mass temperature

There may be additional factors that have the potential to limit productivity. Therefore, according to the need, the indicators described can be supplemented with new ones. One of them is the thermal homogeneity of the melt. In the case of thermal homogeneity, little equipment is instrumented to measure it, and although thermographic cameras can be used to measure the melt temperature distribution in the flow cross-section at the header outlet (as shown in Figure 4.49), they are rarely available in plants.

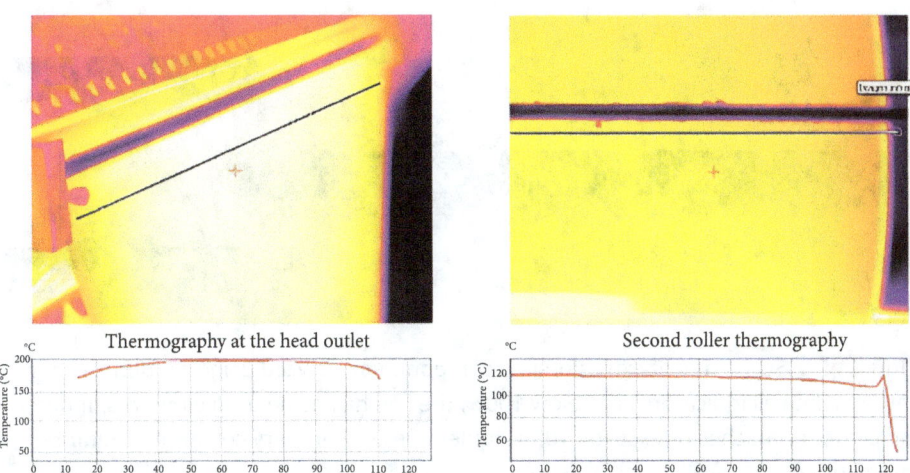

Figure 4.49 Thermographic profile of melt temperature at the die exit of a HIPS sheet extruder. The image highlights areas of non-uniformity in temperature distribution, which can compromise product quality and process control, especially at higher screw speeds

This variable is important because regulating the flow for product calibration is challenging when the mass temperature is not sufficiently homogeneous. Moreover, these differences grow dramatically with increasing screw rotation speed. The factor associated with mass temperature variations can be calculated with the following equation:

$$f_{\sigma_{\Delta T_m}} = \frac{\sigma_{\Delta T_m} \; [^\circ C]}{\sigma_{\Delta T_{m,max}} \; [^\circ C]} \tag{4.56}$$

The maximum standard deviation of the mass temperature ($\sigma_{\Delta T_{m,max}}$) depends on the quality requirements of the product. More permissive products allow processes with higher variations. It is known to be a determining factor, but there is little information on recommended values according to the process or type of product. T_m variation problems can be solved through the temperature profile or system restriction. Otherwise, critical technological changes or implementations are required, such as the investment in new screws or the installation of static mixers in the extrusion line.

Figure 4.50 shows an indicator that can be calculated for each process step. When all bottleneck indicators are below 1 (in absolute terms) or 100% (in percentage terms), there is no apparent reason why the line should not be able to produce at higher speed. When any of the indicators get dangerously close to the limits, it becomes the bottleneck, and plans must be put in place to expand capacity or reduce the current value of the variable in the process. From this point on, using the previously discussed methodologies, such as FMEA, the eight disciplines, problem-solving logs, and cause–effect diagrams (among others) can be helpful. In this way, it becomes feasible to increase productivity.

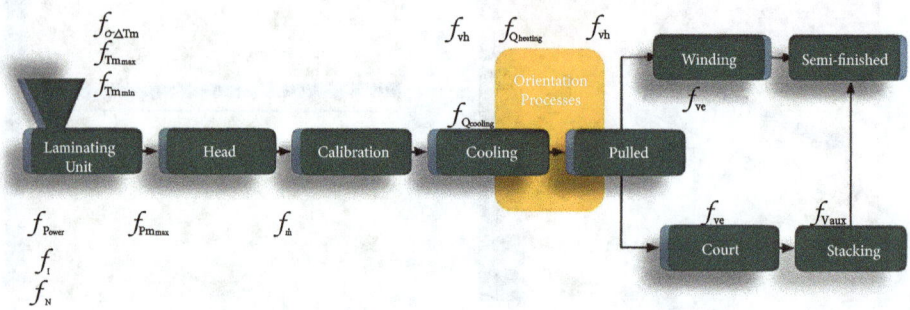

Figure 4.50 Schematic of extrusion line components analyzed using bottleneck indicators. When all indicators are below 100%, the line has no limiting elements. If one indicator approaches or exceeds 100%, it identifies the primary constraint requiring intervention

4.5.1.2 Gap Reduction in Injection Molding: A Process Approach

There are three ways to reduce the production gap in injection molding to lower specific energy consumption:

- Reduction of cycle time: In injection molding, productivity is closely tied to reducing cycle time. The injection molding machine continuously demands energy to activate the motors and maintain the machine and polymer at the required temperature. Consequently, a shorter cycle time results in a more efficient process

- Reducing energy consumption per cycle: After optimizing the cycle time, further actions can be taken to improve energy performance by adjusting processing parameters

- Defining mold and injection molding machine specifications with energy criteria: Energy optimization in the injection molding process is closely related to the proper matching of the machine and the number of cavities defined for a mold.

In the following paragraphs, these three approaches will be discussed in detail.

4.5.1.2.1 Reduction of Cycle Time

A typical measurement of energy consumption during a cycle is presented in Figure 4.51. The process stages are mold closing, filling, packing pressure, cooling, plasticizing, mold opening, and part ejection. The injection molding machine has a constant "base demand", regardless of the cycle activity, to keep the machine and polymer warm. New servo injection machines and all-electric systems have a very low base consumption compared to older hydraulic injection machines. Decreasing cycle time diminishes the impact of this base demand on overall energy consumption, lowering specific energy consumption. This is the primary reason why shorter cycle times generally result in lower specific energy consumption.

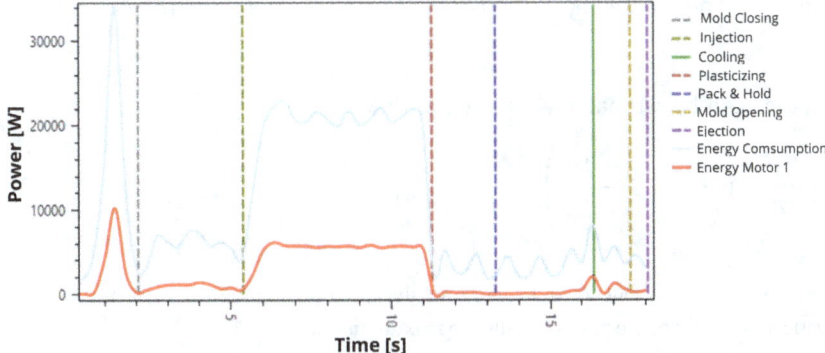

Figure 4.51 Power demands for each stage of the injection cycle for a thermoplastic process. Dotted lines denote the finish time of each stage: mold closing, filling, packing pressure, cooling, plasticizing, mold opening, and part ejection. The red line represents the power consumed by the primary motor, and the blue line represents the total power consumed by the injection machine

Mold and robot movements are apparent targets to reduce cycle time. However, safety considerations to protect the mold and machine should always be a priority. Therefore, the final motion of the mold is usually at a low speed. Typically, most of the cycle time is spent during the cooling or plasticizing stages, making these areas the primary focus for reducing cycle time. Figure 4.52 presents the sequential stages of a typical injection molding process. As shown in the figure, while the mold cools the part, the injection molding machine simultaneously applies the holding pressure and then plasticizes the polymer for the next injection. When the plasticizing and holding pressure time exceeds the cooling time, the process is dominated by plasticizing. Conversely, when the cooling time exceeds the plasticizing and holding pressure time, the process is dominated by cooling.

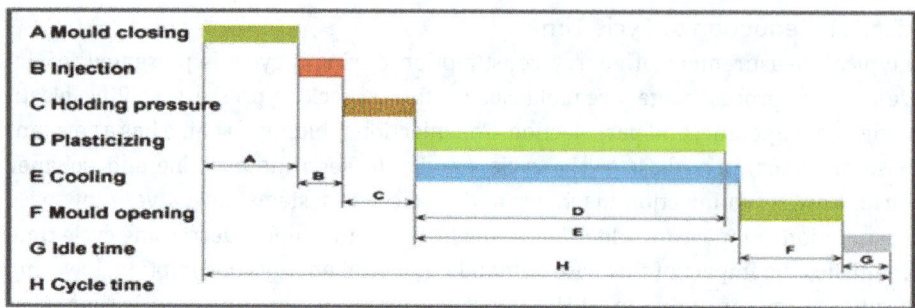

Figure 4.52 Chronological representation of a typical injection molding cycle showing overlapping stages of packing, cooling, and plasticizing. The chart helps identify whether the cycle is dominated by cooling or plasticizing, influencing optimization strategies

The cooling time can be estimated with Equation 4.56:

$$t_{\text{cooling}} = \frac{x_0{}^2}{\alpha\pi^2} \ln\left(\frac{8}{\pi^2} \cdot \frac{\overline{T}_\text{p} - \overline{T}_\text{W}}{\overline{T}_\text{E} - \overline{T}_\text{W}}\right) \qquad\qquad (4.57)$$

where:

t_{cooling} is the time required to cool down the part until the ejection temperature is reached

x_0 is the critical thickness of the part, typically the maximum thickness

α is the thermal diffusivity of the material

$\overline{T}_\text{p}$ is the average temperature of the polymer when it enters the mold

$\overline{T}_\text{W}$ is the average temperature of the mold wall

$\overline{T}_\text{E}$ is the ejection temperature. Note: The ejection temperature is not the surface temperature of the part wall, which is like the mold wall temperature. Instead, it is the average temperature across the thickness of the part, where the polymer achieves sufficient rigidity to prevent deformation during ejection.

The plasticizing time is estimated by dividing the injection weight (calculated as the weight of each part multiplied by the number of cavities plus the weight of the runners) by the machine throughput (typically provided in kg/h). However, machine throughput depends on several factors, including the machine's technology and size, the material's density, enthalpy change, viscosity, and processing conditions. Therefore, the most accurate way to determine the plasticizing time is to measure it during the actual process.

Cooling time is primarily influenced by part thickness: the thicker the part, the longer it takes to cool. Conversely, plasticizing and holding pressure times are determined by the weight injected per cycle; more weight requires a longer time to plasticize the material. Consequently, processes involving parts with relatively small thicknesses and higher weights tend to be dominated by plasticizing. In comparison, processes with thicker parts and lower weights are more likely to be dominated by cooling.

For example, large auto parts such as bumpers or front panels are typically dominated by plasticizing due to their significant weight relative to their thickness. Another example of a process dominated by plasticizing is multi-cavity thin-wall packaging injection molding, where the small thickness results in extremely fast cooling. In the following sections, we will discuss recommendations to reduce the cycle time when it is dominated by plasticizing or cooling.

4.5.1.2.2 Recommendations for Cycle Times Dominated by Plasticizing

We begin with the recommendations related to the technology involved in the process. These recommendations aim to reduce the technological gap, which is explained in detail in Section 4.6.

- When selecting an injection molding machine, prioritize the plasticizing rate or throughput rate, typically measured in units of mass per unit of time

- Avoid using general-purpose screws. Instead, opt for screws specifically designed for the material you will use. This optimization directly impacts the productivity of the process

- Consider using technologies that employ separate mechanisms for holding pressure and plasticizing. Traditional injection molding machines use the same screw for both tasks, which must be performed sequentially. Advanced technologies use an independent piston to apply holding pressure, allowing both operations to occur simultaneously. This reduces cycle times, particularly in processes where plasticizing is the limiting factor

- Be aware that increasing the number of cavities in such cycles does not always lead to higher productivity, as it increases the amount of mass to plasticize. This factor should be carefully considered in the mold design.

Regarding the processing conditions, the following points should be considered:

- **Use the highest rotational speed for the screw that does not cause quality issues.** This parameter is limited by the risk of material degradation due to high shear rates, which can increase the quality gap (discussed in Section 4.4)

- **Use the lowest back-pressure value that does not cause quality issues.** Back-pressure slows the return of the screw during plasticizing to improve the thermal homogeneity of the molten polymer and the quality of the mixture, which is especially important for pigmented materials. However, increased back-pressure extends plasticizing time and demands significant energy. It is recommended to start without back-pressure and gradually increase it only if quality problems arise. To reduce back-pressure, increasing the polymer melt temperature may be beneficial

- **Reduce the duration of the holding pressure.** The impact of holding pressure is influenced by both the applied pressure and its duration. Accurately defining this parameter is essential to ensure the final weight and tolerance of the part. Simulations can help determine the optimal holding pressure parameters, while tools like the design of experiments can aid in using weight and tolerances as dependent variables.

4.5.1.2.3 Recommendations for Cycle Times Dominated by Cooling

As shown in Equation 4.57, cooling is influenced by the part's thickness (part design), the material's diffusivity (material selection), and the temperatures of the ejection, polymer melt temperature, and temperature of the mold walls (processing conditions). These factors are discussed below.

Part design and material selection

An optimized thickness reduces material usage and processing time, making it an engineering challenge best addressed during the design stage. Additionally, using materials with mineral or metallic fillers can significantly increase diffusivity, thus reducing cooling time [33].

Temperature of ejection

Cooling time can be optimized experimentally by testing various durations and identifying the minimum time at which parts can be ejected without quality issues. Alternatively, it can be estimated using Equation 4.57, where the ejection temperature is approximated by the heat deflection temperature (HDT) [34].

Temperature of the mold walls

The secret to a good design in the cooling system of a mold is maintaining a uniform temperature across the entire contact surface with the plastic part throughout the cooling process. However, the complexity of many injected parts makes designing the cooling system highly challenging. Using simulation tools is crucial in optimizing the cooling system design to ensure efficient cycle time and that the final part meets all tolerance and quality specifications [35]. Another tool uses metal additive manufacturing to create conformal cooling channels with complex geometries, allowing for a uniform temperature in the mold walls [36].

Simulation, especially during the design stage, along with capturing thermographic images during mold operations, allows identification of hot spots in the mold. This information can help modify the cooling system or incorporate beryllium copper or other metallic inserts with high thermal conductivity to reduce the temperature in these hot spots. The mold temperature should balance the quality of the part and the cycle time. Typically, higher mold temperatures are needed to improve surface quality, as they increase gloss levels, but this also extends the cycle time. Additionally, higher mold temperatures reduce thermal stresses, minimize post-crystallization deformations, and facilitate the flow of highly viscous polymers in thin parts. In processes dominated by cooling, the mold temperature should be set as low as possible without compromising the quality specifications of the final part. However, always avoid setting the coolant temperature below the dew point. If the mold temperature falls below the dew point, water from the air will condense on the mold surface. This condensation not only wastes the energy used to cool the mold but can also cause damage to the mold and lead to quality issues in the final part.

Cooling efficiency depends on two factors: the temperature of the coolant and its volumetric flow (Reynolds number). A good practice in injection molding and other discontinuous processes, such as blow molding, is to constantly monitor the coolant flow and the temperature difference between the inlet and outlet. High-temperature gradients between the inlet and outlet (5 °C or more) may indicate an insufficient flow rate or a restriction in the cooling channel, suggesting that maintenance is required.

Temperature of the melt

The melt temperature should be set as low as possible to reduce cycle time in cooling-driven processes. This adjustment involves a trade-off between quality and productivity like the mold wall temperature. Lower melt temperatures may increase injection pressure, result in incomplete fillings, and cause undesirable orientations and residual stress.

Plasticization time

Ensure that the plasticization time is matched to be slightly less than the cooling time. Rapid screw retraction and plasticizing use more energy than gentle and slow screw retraction. The screw-back time should be approximately 80–90% of the cooling time. However, this may be difficult to achieve if the only control action available is the screw rotational speed.

4.5.1.2.4 Reduction of Energy Consumption per Cycle

Energy consumption per cycle can be reduced by lowering the overall energy demand or minimizing energy consumption time. The latter strategy was discussed in the previous section. To address the former, monitoring energy consumption and identifying energy usage at each cycle stage is essential, prioritizing stages with higher consumption.

If plasticizing is a significant factor, conducting a series of design experiments is advisable. This will help optimize energy consumption by balancing the plasticization velocity and temperature without impacting cycle time. Additionally, using the lowest back-pressure value that does not compromise quality is recommended.

One measure that could help reduce the technological gap (Section 4.6) and lower the energy consumption per cycle is a toggle clamping system. These consume less energy for long cooling times than hydraulic clamping systems, which require pumps to run continuously to maintain pressure while the mold is closed. In contrast, toggle clamping systems only consume energy during the clamping and unclamping operations. Once the mold is closed, the toggle mechanism locks in place without requiring additional energy to sustain the closing pressure.

Another measure is to use variable-frequency drives. In some machines, most of the energy expended is due to the operation of the motors, even when they are not performing any tasks. To reduce this consumption, using variable frequency drives is highly desirable, as they allow the regulation of motor speed based on the required task. This adjustment increases the overall efficiency of the process.

4.5.1.2.5 Defining Mold and Injection Molding Machine Specifications with Energy Criteria

The energy performance of the injection molding process is limited by the initial decisions made during the mold design and the specification definition of the injection molding machine required for the project. An adequate match between the mold and injection molding specifications is key to obtaining higher efficiency. Many specifications, such as the clamping force required and the mold size, are relevant to selecting a machine for a mold or designing a mold. However, from the energy point of view, we will concentrate this definition on the maximum mass that the machine can plasticize and the mass injected per shot into the mold, which is a function of the number of cavities.

As we discussed, the more mass that is molded per shot, the higher the energy efficiency, mainly if the cycle time is driven by cooling. If the cycle time is driven by plasticization, this hypothesis is not always accurate, and other considerations should be made.

The plasticization capacity per shot is usually a function of the screw diameter, as follows:

$$W = \frac{l \cdot \pi \cdot D^2 \cdot \rho}{4} \tag{4.58}$$

where:
W is the weight of the shot given [g]
D is the diameter of the screw
ρ is the density of the polymer at the temperature of plasticization
l is the axial movement of the screw during plastification.

As shown in Figure 4.53, the value of l should be between $1D$ and $3D$, and only under special conditions up to $4D$. When values lower than $1D$ are used, the machine's movements are not accurate enough to guarantee consistent parts, and the injection molding machine becomes too large for the mold, resulting in low energy-efficiency. Conversely, when l exceeds $4D$, quality problems may arise, such as air in the melt, leading to defects in the final part.

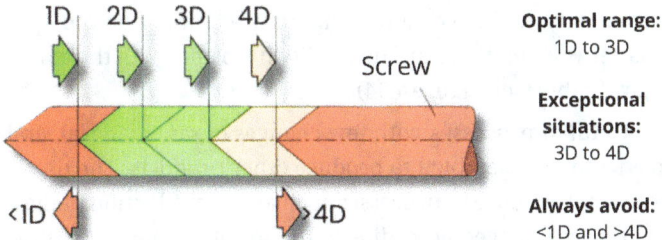

Figure 4.53 Recommended axial screw travel ranges for injection molding machines. The ideal range is between 1*D* and 3*D*, ensuring process stability and part consistency. Values below or above this range may compromise energy-efficiency or product quality

Under this hypothesis, the more mass molded per shot, the more energy-efficient the process is, and a value of $l = 3D$ per shot should be targeted. Therefore, to calculate the ideal number of cavities from the energy performance point of view, the following equation can be used:

$$n = \frac{3 \cdot \pi \cdot D^3 \cdot \rho}{4 \cdot w} \tag{4.59}$$

where:

n is the ideal number of cavities

w is the weight per part, including the runner system in a cold runner mold.

On the other hand, if the mold already exists, the suggested diameter of the injection molding machine can be estimated as follows:

$$D = \sqrt[3]{\frac{4 \cdot W}{3 \cdot \pi \cdot \rho}} \tag{4.60}$$

Equation 4.59 and Equation 4.60 are suggested values based on the maximum utilization of the mass molded per shot. However, other relevant variables should be considered, such as the investment cost of additional cavities in the mold, the size of the mold as a function of the number of cavities, and the clamping force that is a function of the number of cavities, among others.

4.5.2 Extrusion Operating Curves

The operating curve is a recommended tool for the extrusion process. The operating curve relates the screw rotation speed with the restriction generated by the die to determine the operating points and the equipment response in mass flow and melt pressure. With the proper execution of the procedure, it is also possible to calculate the

limits of profitability, homogeneity, and maximum temperature that define the optimum operating area. The latter is when there are no quality problems with the extrudate, and the operation is profitable (see Figure 4.54).

Operating curves are very useful in processes with few changes of raw material, and
different dies are assembled and disassembled to produce other products. The determination of operating curves makes sense in processes such as the production of plastic pipe or profiles, where dies are changed according to the specifications of the pipe
or profile to be produced. They also have value in the extrusion blow-molding of hollow bodies, where there are dies for each product, and in the extrusion of the sheet
for thermoforming, where the flexible lips of the extrusion die are closed or opened
according to the thickness of the sheet to be produced.

It is worth noting that the operating curves provide information on the operation of
the plasticizing unit–die assembly but do not provide information on the post-extrusion equipment, which are all the components downstream of the die. Therefore, the
operating curve does not provide information on whether production limitations exist in any of these components. In Figure 4.54, the lines marked k_1 to k_6 are the die's
restriction lines or characteristic lines. A restriction line is characterized by a die
(nozzle + distributor) and the screen pack selected for melt filtration. Any change in
the machine configuration from the point where the screw ends modifies the restriction. The lines marked n_1 to n_6 are different screw rotation speeds and are known as
screw characteristic lines. The intersection between a screw characteristic line and a
screw characteristic line is called the operating point.

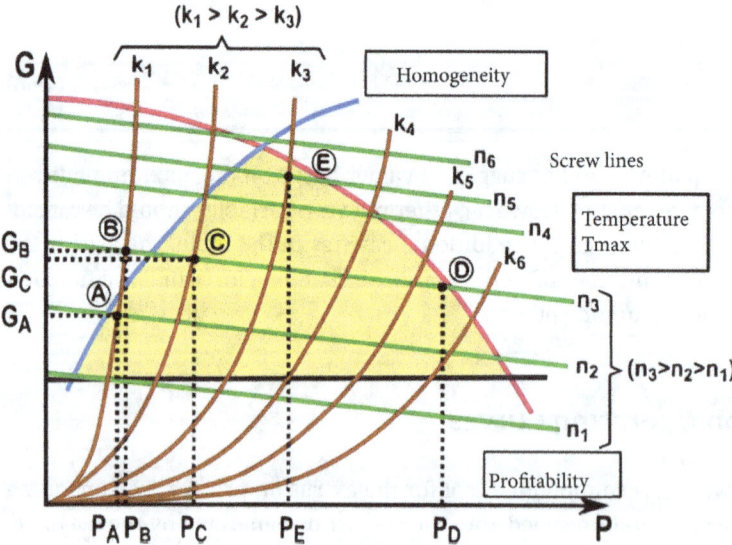

Figure 4.54 Experimental operating curve of an extruder showing melt pressure vs. mass
flow for different screw speeds and die restrictions. The chart defines optimal processing
zones by identifying limits imposed by temperature, homogeneity, and profitability

The operating curve should be constructed by following this experimental procedure:

1. Select the raw material

2. Adjust the temperature profile suitable for the machine and the material used

3. Select the constraints or restrictions to be mounted. It is essential that at least three different restrictions can be fitted

4. Select the screw speeds to be used that cover the range up to the maximum rotational speed. Usually, four to six different rotational speeds are selected. If three restrictions and five screw rotation speeds were selected, 15 operating points would be evaluated

5. Perform the assembly of the first restriction

6. Adjust the first rotation speed

7. Allow the process to stabilize for at least 10 min

8. Measure the mass flow, melt temperature, and melt pressure, and report it in a format like the one presented in Table 4.12

9. Repeat the procedure with a new rotational speed from step 7, until the selected maximum speed is reached

10. Perform the assembly of the following restriction

11. Repeat the procedure from step 6 until all the constraints are completed

12. Create a Cartesian coordinate plot with mass flow on the y-axis and melt pressure on the x-axis. Plot $n \times m$ operating points on the graph, where n is the number of rotational speeds and m is the number of constraints or restrictions in determining the operating curve

13. Join the points with the same constraint and identify them with the constraint that characterizes it

14. Draw a trend line through the points having the same rotational speed. These lines must be parallel. Identify it with the rotation speed that characterizes it

15. Document the machine, the material, and the temperature profile used in the operating curve.

Table 4.12 An Example of a Format for Documenting Operating Conditions for Constructing the Operating Curve. The Blank Cells Must be Fulfilled for Each Operating Condition

Restriction	Rotation speed [rpm]	Mass flow rate [kg/h]	Melt pressure [bar]	Melt temperature [°C]
k_1	n_1			
...	...			
k_n	n_n			

The profitability line is determined by establishing all fixed costs and answering the question: What must be the minimum productivity to pay the fixed costs? A horizontal line is drawn at the value of this mass flow. The maximum temperature and homogeneity limits are more complex and time-consuming to determine. Still, to find them, the experience of the plant with the machine is transferred to the diagram or operating curve. For each product, it is necessary to work at the highest operating point in the diagram. The more restricted the head, the greater the chance of quality problems due to overheating the melt. The less constrained the head, the more likely quality problems associated with melt homogeneity will appear. If the problematic operating points are marked on the operating diagram, the optimum extruder operating area for the material and temperature profile conditions employed will be defined over time. This methodology can complement the bottleneck method, as it includes melt quality aspects.

4.5.3 Design of Experiments

This methodology is especially recommended for the thermoplastic injection process due to the significant number of control variables. The design of experiments allows a response surface to be obtained that can be used to predict the behavior of the process to any condition within the variation ranges chosen for the variables of interest (interpolation) or slightly outside them (extrapolation). This response surface allows the conditions that optimize the response variable, usually the cycle time, to be determined.

4.5.4 Process Modeling and Simulation

The development of plasticizing units for thermoplastics processing aims to make the equipment increasingly more productive and stable. Also, it can produce high-quality products with an increasingly rational and efficient energy use. All this is to meet the need to increase industrial competitiveness. Technological development in polymer processing and the ability to model the physical phenomena that explain the behavior of technology and allow for its optimization, improvement, and exploitation are outdated. Technology usually breaks through before the models. Therefore, understanding the mechanisms that make technology work and make it work with higher productivity, greater energy efficiency, or improved product quality is a necessary step to refine the technologies developed. Models are what make process simulation possible.

Process simulation allows modification of processing conditions and estimation of the result from the solution of the phenomenological models that explain its behavior. Among the computational tools that can be used are ExtruTools and InyecTools from the Instituto de Capacitación del Plástico y del Caucho (ICIPC). These use analytical models and are based on numerical methods such as finite volumes, finite elements, boundary elements, and the technique of radial functions. Among the latter, programs for the simulation of mold filling in the injection molding process are among the most useful, such as CADMOULD (see Figure 4.55), MOLDFLOW, and MOLDEX, among others. In all cases, knowledge of the behavior of the properties of plastic materials and their relationship with the different transformation processes is fundamental.

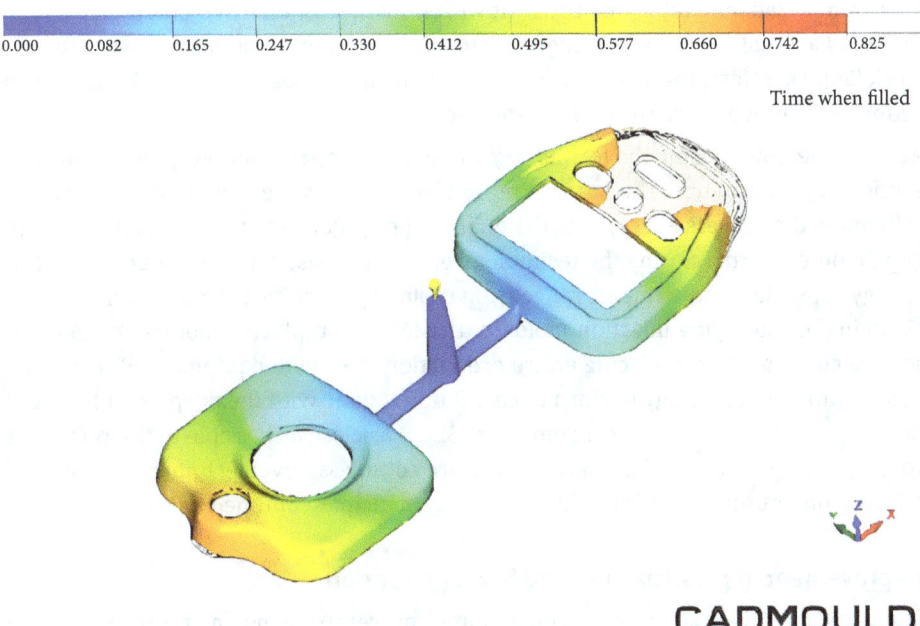

Figure 4.55 Simulation of mold filling using CADMOULD software, demonstrating polymer flow behavior during injection molding. Such tools support optimization of gate location, injection time, and fill balance, contributing to energy and material efficiency

4.6 Closing the Technology Energy Gap and the R&D Energy Gap

The energy efficiency of an industrial process is strongly related to the technological level of the production line. Both the technology energy gap and the R&D energy gap are associated with improving the technology used for processing. The difference is that the technology energy gap is a gap that is under the control of the factories and represents the relative distance (from an energy point of view) between the technologies used and the best technologies available in the state of the art and which are commercial technologies. The research & development & innovation (R&D) energy gap is reduced to the extent that investment is made in developing new technologies, which is why it usually points to technology not yet available on the market. For polymer processing plants, the technology energy gap is the costliest to close as it requires significant investments. Likewise, the R&D energy gap is the most expensive for machinery manufacturers to close. As such, they are by far the most difficult decisions to make. Typically, investment decisions are based on payback analyses. Anderson & Newell [37] analyzed technology adoption decisions, and as expected, the adoption rate is higher for investments with short amortizations and lower costs. However, many cost-effective energy efficiency projects have not been implemented due to a lack of an internal management framework [38]. Therefore, investment decision methodologies must be combined with other actions to improve energy efficiency effectively [39].

At the same time, closing the technology gap usually impacts energy, production, and economic performance. The latter is more significant, as the technology energy gap influences the process energy gap, the quality energy gap, and the production energy gap. In other words, closing the technology energy gap also helps to reduce the other energy gaps. These investments range from replacing machine parts and components, changing or modifying injection molds, extrusion dies replacing motors and general service equipment, to replacing entire production lines with equipment of higher efficiency and better energy performance. All this is done with the purpose of bringing the machine specific energy consumption (SEC_m) as close as possible to the reference specific energy consumption (SEC_b). There are countless novel technologies, but they all work on improvement from the same basic principles, outlined below.

Improvement of plasticization and homogenization

The most energy-consuming thermodynamic processes in polymer processing are plasticization and heating the polymer to the appropriate process temperature. Temperature homogenization and melt composition, on the other hand, allow the plasticizing speed of the processing equipment to be exploited. Since the advent of synthetic plastics, work has also been done on the technological improvement of plasticizing units to make them increasingly efficient and productive. Figure 4.56 shows the evolution of single-screw plasticizing units for polymer processing. As can be seen in the figure, the technology of single-screw plasticizing units has been developing towards longer units, integrating the barrier flight with distributive and dispersive mixing zones of different geometries and higher intensity. There are also units with grooved feed zone (GFE) and then grooved plasticizing zone (GPE) or high-speed grooved feed zone (HSSSE) units [40]. In this evolution, single-screw plasticizing units increased their productivity and energy efficiency, which allowed them to considerably reduce SEC_m.

Both the GPE and HSSSE plasticizing units, with 75 mm diameter screws and lengths varying between 36D and 40D, can produce 600–1000 kg/h using polyolefins, with an SEC_m as low as 0.18–0.22 kWh/kg. For reference, a conventional extruder of the same characteristics and with the same material produces 150–350 kg/h with an SEC_m of 0.26–0.46 kWh/kg [39]. This is 3–4 times more productive than a conventional extruder and 2 times more than a GFE, with a specific energy consumption up to half that of its counterparts [7].

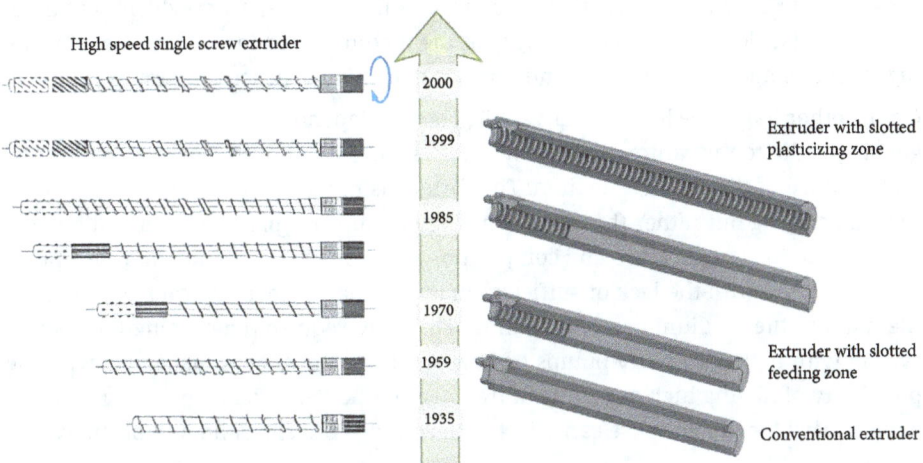

Figure 4.56 Technological evolution timeline of single-screw plasticizing units for polymer processing. Advancements include the integration of mixing elements and grooved zones to improve melt homogenization, throughput, and energy efficiency

In addition, there are many other plasticizing unit technologies: the one-rotation-per-oscillation oscillating single-screw plasticizing unit or Buss kneader [41]; the three-rotation-per-oscillation oscillating single-screw plasticizing unit or Trivolution [42]; the co-rotating, counter-rotating, conical and cylindrical twin-screw plasticizing units [43, 44]; parallel [45] and delta [46] triple-screw plasticizing units; planetary gear plasticizing units [47]; planetary screw plasticizing units [48]; and screwless plasticizing units using a rotor-stator system, known as CONEX [49], among others. These technologies are intended to show all the technological variations found on the market. In addition, they illustrate how quickly new technologies appear that make others obsolete or solve specific production problems, opening a niche where these technologies represent a better solution than traditional options.

Many examples of new technologies are highly innovative, efficient, and productive. The problem with referencing them in this book is that technology becomes outdated quickly, so this chapter would be as ephemeral as the speed with which new technologies appear on the market. Therefore, from this point on, specific technologies will no longer be referenced.

Improvement of cooling

Heating and cooling are required to process a polymer. The latter consumes almost as much energy as the former. Some processes and technologies are limited by plasticizing capacity and others by cooling capacity. Cooling must meet several characteristics: it must be fast, homogeneous, and efficient. It is usually performed with air or water or in contact with a cooled surface. The cooling rate is associated with the temperature gradient between the cooling medium and the polymer from which heat is removed. It also depends on the flow rate of the medium used for cooling. Usually, the higher the temperature gradient and flow rate, the higher the cooling rate.

On the other hand, the homogeneity of the cooling depends on two factors: the homogeneity of the temperature of the polymer to be cooled and the homogeneity of the temperature of the cooling medium. The former is not a problem that can be solved through cooling but rather through technology to improve plasticization and homogenization of the temperature and composition of the melt. The second may be a problem associated with the lack of sufficient mass for the hot mass to transfer energy to the mass of the medium used for cooling. This may require redesigning the cooling system with higher-capacity pumps or fans. Cooling efficiency is associated with the possibility of having high overall heat transfer coefficients. The type or design of the devices used for this purpose can affect both cooling efficiency and homogeneity.

In this sense, in injection molding, blow molding, and thermoforming processes, the proper design of mold cooling channels and water flows that ensure sufficiently high Reynolds numbers to maximize heat transfer coefficients is required. Sometimes, reaching specific areas of the part with cooling channels is difficult. This is how new technologies appear for mold cooling, such as conformal cooling, which deals with cooling channels of quite complex geometries obtained by 3D printing of metals [50]. There is also the use of heat pipes [51], which are devices of infinite thermal conductivity; they can be so thin that they can reach regions of the part where it is impossible to carry a machined cooling channel.

In extrusion, there are a wide variety of cooling technologies, each appropriate and adequate to the type of process. For this reason, it is imperative to maintain a continuous technological watch. Processes inherently limited by cooling, such as blown film extrusion, constantly evolve in response to the evolution of the technological devices affecting this part of the process. From single-lip cooling rings, it evolved to the use of double-lip rings. Subsequently, internal bubble cooling technology appeared. Then, blown film technologies with triple-lip and double-cooling rings appeared [52]. There is now blown film equipment with water ring cooling [53].

Improvement of the cleaning times required during material and color changes

Technological implementations, in this sense, seek additionally to reduce the production energy gap and have a significant effect on the reduction of the base load. Material and color changes become more complex and time-consuming with high residence time, when the technology has a wide distribution of polymer residence times, or when the polymer flow occurs through channels with many direction changes without a smooth transition between them. To reduce residence times, technologies seek to minimize flow channel path length and volume inside plasticizing units and dies. The plasticizing unit technologies that offer a narrower distribution of residence times are plasticizing units with a bigger component of elongational flow than shear flow. In this sense, conventional single-screw plasticizing units usually have only a shear flow component and have the widest residence time distributions. In this sense, changing color and material in this type of technology is more complex and time-consuming.

An example is the intervention of a polypropylene multifilament extrusion line. The line had a 76.2 mm diameter extruder with a length of 34D that fed four rows, each producing slightly more than 60 filaments. The company makes more than 200 color changes monthly, generating a fiber rejection rate of 25–35% due to color deficiencies. The extruder ran at 48–68 rpm with a mass flow rate of 65–90 kg/h. The system had the configuration shown in Figure 4.57(a).

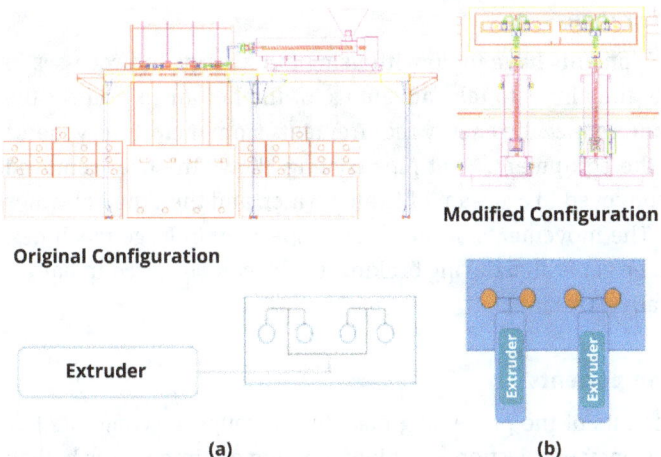

Figure 4.57 Configuration of a polypropylene multifilament extrusion line before (a) and after (b) technological intervention. The modified setup improves flow distribution, reduces color change time, and minimizes material degradation, enhancing both quality and productivity

To increase the versatility of the system, the original plasticizing unit was replaced with two 45 mm diameter, 25D long plasticizing units, each feeding two dies. The plasticizing units were strategically placed to minimize the distance between the plasticizing unit and the dies and to reduce directional changes, as shown in Figure 4.57 (b). With this configuration, the channel lengths are significantly reduced (about 80%), and the changes in direction fall from 10 to 2. In addition, the diameters of the melt conduction pipes between the plasticizing unit and the windrow were reduced. The diameters were calculated rheologically. Thus, the multifilament extrusion line was able to produce 35–90 kg/h at screw rotation speeds of 43–200 rpm. It was also able to produce two colors simultaneously. With these modifications, the line reduced the rejection rate to

2.2–5% per month. Additionally, the shorter residence time reduced the natural degradation of the material, which lowered the DPF (denier per fiber) from 3.33 to 2.08, maintaining the tenacity of the yarn. In addition, the yarns obtained allowed high spinning speeds previously only possible with polyester yarns.

Reduction of the time required for reference changes, and automating manual processes and tasks

In principle, this has the same purpose of further reducing the production energy gap and the base load but with another type of technology. In this case, these are devices or complements that automate and accelerate processes and tasks, such as quick mold change systems, quick hopper cleaning systems, quick material change, and autotuning for adjusting operating conditions, among others.

Use of energy lost in the process

Many technological developments have to do with using lost energy. In processes in which air cooling is used, the aim is to take advantage of the hot air to preheat the polymer in the hoppers and reduce the energy requirements from the primary motor or the heating bands of the equipment. Heat pipe cooling allows the heat removed from the polymer to be conveyed to a stream of heated water, and then the hot water is used within the plant. The movements of the molds, especially in large machines, allow electrical energy to be generated during braking, which can be stored in batteries for later use (regenerative braking).

Use of more efficient components

The control of the movements of the processing machines through servomotors has increased the energy efficiency of injection and blow molding equipment for hollow bodies. Using more energy-efficient electric motors in extrusion, some of which do not require a speed reduction system, and heating and temperature control systems that minimize energy losses to the environment and reduce thermal inertia, have had the same impact on the operation of plasticizing units. Ultimately, achieving a highly energy-efficient system requires efforts on several fronts, to make each of the stages and components of the process more efficient.

Implementing online quality control

This is one of the most important trends in recent times. It involves moving much of the quality control from the laboratory to the processing equipment so that quality control occurs in line with production. This has been made possible by the accelerated development of optical inspection systems and image processing supported by AI. In-line measurement of thickness, width, diameter, ovality, shape, and determination of the presence of contaminants and surface or appearance defects of the product is performed. In addition, the measurement through optical tools of color, transparency, opacity, and even the correlation of optical density with properties such as permeation has also been implemented [54, 55]. Determination of the quality of the feedstock being fed in terms of particle size and shape, packing density, and the presence of contaminants in the feedstock is already a reality today [56]. Determining the nature of polymers online at high processing speeds has been achieved using near-infrared (NIR) detectors, medium infrared (MIR) detectors, hyperspectral imaging, and laser-induced breakdown spectroscopy (LIBS). This is an area of constant and accelerated evolution.

Implementing process integration

Process integration is another of the most significant technological trends of recent years. Process integration seeks to obtain practically finished products in a single step or from a single technology, minimizing energy and personnel requirements.

Keeping up to date with new technologies

For all the reasons mentioned above, companies should establish an internal technology watch procedure with access to patent databases, scientific articles, and gray literature. In addition, they should frequently visit the various technology fairs in their business area to keep abreast of new technological developments and plan the closing of the organization's technological gap. With these actions, we can learn about the efforts made regarding equipment, parts, component, and machinery manufacturers, and use this information to reduce the R&D energy gap, which, if not addressed, will also widen the technology gap for companies in the future.

References

[1]   O. Estrada, I. D. López, A. Hernández, J. C. Ortíz, "Energy gap method (EGM) to increase energy efficiency in industrial processes: Successful cases in polymer processing", *Journal of Cleaner Production*, 2018, vol. 176, pp. 7–25, DOI: 10.1016/j.jclepro.2017.12.009

[2]   R. D. Peterson and C. K. Belt, "Elements of an energy management program", *JOM*, 2009, vol. 61, pp. 19–24, DOI: 10.1007/s11837-009-0046-2

[3]   "EUROMAP 60.1: Injection moulding machines – Determination of machine related energy efficiency class" [Technical Recommendation 60.1], EUROMAP, 2013, *https://www.euromap.org/media/recommendations/60/2013/EU_60.1_Jan_2013.pdf*

[4]   "EUROMAP 46.1: Extrusion blow moulding machines – Determination of machine related energy efficiency class" [Technical Recommendation 46.1], EUROMAP, 2014, *https://www.euromap.org/media/recommendations/60/2013/EU_60.1_Jan_2013.pdf*

[5]   E. Hirst and M. Brown, "Closing the efficiency gap: Barriers to the efficient use of energy", *Resources, Conservation and Recycling*, 1990, vol. 3, pp. 267–281, DOI: 10.1016/0921-3449(90)90023-W

[6]   M. Schulze, H. Nehler, M. Ottosson, P. Thollander, "Energy management in industry – A systematic review of previous findings and an integrative conceptual framework", *Journal of Cleaner Production*, 2016, vol. 112, pp. 3692–3708, DOI: 10.1016/j.jclepro.2015.06.060

[7]   O. Estrada, J. C. Ortiz, A. Hernández, I. López, F. Chejne, M. del Pilar Noriega, "Experimental study of energy performance of grooved feed and grooved plasticating single screw extrusion processes

in terms of SEC, theoretical maximum energy efficiency and relative energy efficiency", *Energy*, 2020, vol. 194, article no. 116879, DOI: 10.1016/j.energy.2019.116879

[8]   H. Erbiyik, "Definition of maintenance and maintenance types with due care on preventive maintenance", in *Maintenance Management – Current Challenges, New Developments, and Future Directions*, G. Lambert-Torres, E. Leandro Bonaldi, L. Eli De Lacerda De Oliveira (Eds.), IntechOpen, 2023, DOI: 10.5772/intechopen.106346

[9]   R. Kee, "The sufficiency of product and variable costs for production-related decisions when economies of scope are present", *International Journal of Production Economics*, 2008, vol. 114, pp. 682–696, DOI: 10.1016/j.ijpe.2008.03.003

[10] I. B. Da Silva and M. Godinho Filho, "Single-minute exchange of die (SMED): A state-of-the-art literature review", *The International Journal of Advanced Manufacturing Technology*, 2019, vol. 102, pp. 4289–4307, DOI: 10.1007/s00170-019-03484-w

[11] S. Muotka, A. Togiani, J. Varis, "A design thinking approach: Applying 5S methodology effectively in an industrial work environment", *Procedia CIRP*, 2023, vol. 119, pp. 363–370, DOI: 10.1016/j. procir.2023.03.103

[12] B. Strauch, *Investigating Human Error*, CRC Press, 2017, DOI: 10.1201/9781315589749

[13] E. R. Sykes, "Interruptions in the workplace: A case study to reduce their effects", *International Journal of Information Management*, 2011, vol. 31, pp. 385–394, DOI: 10.1016/j.ijinfomgt.2010.10.010

[14] S. Iwao, "Revisiting the existing notion of continuous improvement (Kaizen): Literature review and field research of Toyota from a perspective of innovation", *Evolutionary and Institutional Economics*, 2017, vol. 14, pp. 29–59, DOI: 10.1007/s40844-017-0067-4

[15] D. E. Embrey, "SHERPA: A systematic human error reduction and prediction approach", in *Proceedings of the International Topical Meeting on Advances in Human Factors in Nuclear Power Systems*, American Nuclear Society, 1986, pp. 184–193, *https://inis.iaea.org/records/deen4-r5y37*

[16] Y. Torres-Medina, "El análisis del error humano en la manufactura: Un elemento clave para mejorar la calidad de la producción", *Revista UIS Ingenierías*, 2020, vol. 19, pp. 53–62, DOI: 10.18273/revuin. v19n4-2020005

[17] C. Singh, D. Singh, J. S. Khamba, "Exploring an alignment of lean practices on the health and safety of workers in manufacturing industries", *Materials Today: Proceedings*, 2021, vol. 47, pp. 6696–6700, DOI: 10.1016/j.matpr.2021.05.116

[18] T. A. Saurin, J. L. D. Ribeiro, G. Vidor, "A framework for assessing poka-yoke devices", *Journal of Manufacturing Systems*, 2012, vol. 31, pp. 358–366, DOI: 10.1016/j.jmsy.2012.04.001

[19] C. A. Naranjo, E. M. del Pilar Noriega, M. J. D. Sierra, J. R. Sanz, Injection Molding Processing Data [2nd edition] ("Plastics Pocket Power" series), Hanser Publications, 2019

[20] C. A. Naranjo (Ed.), Extrusion Processing Data ("Plastics Pocket Power" series), Hanser Publications, 2001

[21] A. J. J. Braaksma, A. J. Meesters, W. Klingenberg, C. Hicks, "A quantitative method for failure mode and effects analysis", *International Journal of Production Research*, 2012, vol. 50, pp. 6904–6917, DOI: 10.1080/00207543.2011.632386

[22] R. J. Kent, *Quality Management in Plastics Processing: Strategies, Targets, Techniques and Tools*, Elsevier, 2016

[23] A. Ruiz-Falcó Rojas, "Control estadístico de procesos", Universidad Pontificia de Comillas, 2006, *https://web.cortland.edu/matresearch/controlprocesos.pdf*

[24] S. Bisgaard, "Quality management and Juran's legacy", *Quality Engineering*, 2008, vol. 20, pp. 390–401, DOI: 10.1080/08982110802317398

[25] S. S. Raj, K. A. Michailovich, K. Subramanian, S. Sathiamoorthyi, K. T. Kandasamy, "Philosophy of selecting ASTM standards for mechanical characterization of polymers and polymer composites", *Materiale Plastice*, 2021, vol. 58, pp. 247–256, DOI: 10.37358/MP.21.3.5523

[26] B. Sarkar and S. Saren, "Product inspection policy for an imperfect production system with inspection errors and warranty cost", *European Journal of Operational Research*, 2016, vol. 248, pp. 263–271, DOI: 10.1016/j.ejor.2015.06.021

[27] C. W. Kang, M. B. Ramzan, B. Sarkar, M. Imran, "Effect of inspection performance in smart manufacturing system based on human quality control system", *The International Journal of Advanced Manufacturing Technology*, 2018, vol. 94, pp. 4351–4364, DOI: 10.1007/s00170-017-1069-4

[28] H. Rendón, *Control Estadístico de Calidad* [1st edition], Centro Editorial de la Facultad de Minas, 2013, *https://minas.medellin.unal.edu.co/centro-editorial/libros/control-estadistico-de-calidad*

[29] K. C. Wong, K. Z. Woo, K. H. Woo, "Ishikawa diagram", in *Quality Improvement in Behavioral Health*, W. O'Donohue and A. Maragakis (Eds.), Springer International Publishing, 2016, pp. 119–132, DOI: 10.1007/978-3-319-26209-3_9

[30] L. Liliana, "A new model of Ishikawa diagram for quality assessment", *IOP Conference Series: Materials Science and Engineering*, 2016, vol. 161, article no. 012099, DOI: 10.1088/1757-899X/161/1/012099

[31] A. Grosfeld-Nir, B. Ronen, N. Kozlovsky, "The Pareto managerial principle: When does it apply?", *International Journal of Production Research*, 2007, vol. 45, pp. 2317–2325, DOI: 10.1080/00207540600818203

[32] K. M. Cantor and P. Watts, "Plastics processing", in *Applied Plastics Engineering Handbook*, Elsevier, 2011, pp. 195–203. DOI: 10.1016/B978-1-4377-3514-7.10012-1

[33] B. Weidenfeller, M. Höfer, F. R. Schilling, "Thermal conductivity, thermal diffusivity, and specific heat capacity of particle filled polypropylene", *Composites Part A: Applied Science and Manufacturing*, 2004, vol. 35, pp. 423–429, DOI: 10.1016/j.compositesa.2003.11.005

[34] C. J. Yu and J. E. Sunderland, "Determination of ejection temperature and cooling time in injection molding", *Polymer Engineering & Science*, 1992, vol. 32, pp. 191–197, DOI: 10.1002/pen.760320305

[35] C. Fernandes, A. J. Pontes, J. C. Viana, A. Gaspar-Cunha, "Using multi-objective evolutionary algorithms for optimization of the cooling system in polymer injection molding", *International Polymer Processing*, 2012, vol. 27, pp. 213–223, DOI: 10.3139/217.2511

[36] H. M. Silva, J. T. Noversa, L. Fernandes, H. L. Rodrigues, A. J. Pontes, "Design, simulation and optimization of conformal cooling channels in injection molds: A review", *The International Journal of Advanced Manufacturing Technology*, 2022, vol. 120, pp. 4291–4305, DOI: 10.1007/s00170-022-08693-4

[37] S. T. Anderson and R. G. Newell, "Information programs for technology adoption: The case of energy-efficiency audits", *Resource and Energy Economics*, 2004, vol. 26, pp. 27–50, DOI: 10.1016/j.reseneeco.2003.07.001

[38] S. Aflaki, P. R. Kleindorfer, V. S. deMiera Polvorinos, "Finding and implementing energy efficiency projects in industrial facilities", *Production and Operations Management*, 2013, vol. 22, pp. 503–517, DOI: 10.1111/j.1937-5956.2012.01377.x

[39] O. A. Estrada Ramírez, "Estudio de la influencia del proceso de plastificación en la eficiencia energética del proceso de extrusión monohusillo" [Ph.D. thesis], Universidad Nacional de Colombia, 2021, *https://repositorio.unal.edu.co/handle/unal/79375*

[40] O. Estrada and F. Chejne Janna, "A novel melting model for polymer extrusion: Mechanically induced transition layer removal", *Polymer Engineering & Science*, 2022, vol. 62, pp. 3290–3309, DOI: 10.1002/pen.26104

[41] "Mixing and kneading technologies", BUSS, *https://busscorp.com/* [accessed 3 November 2024]

[42] "Reciprocating continuous kneaders – Tri-kneader mixer", B&P Littleford, *https://www.bplittleford.com/trivolution.html* [accessed 24 March 2025]

[43] "Leistritz extrusion technology" (Twin screw extruders & systems), Leistritz Extrusionstechnik GmbH, *https://extruders.leistritz.com/en/extruders-systems/overview* [accessed 24 March 2025]

[44] "Extruders & compounding machines" (Products and Services), COPERION, *https://coperion.com/en/products-services/extruders-compounding-machines* [accessed 24 March 2025]

[45] "Triple screw extruder, SAT-T Parallel Compounder Manufacturer", USEON, *https://www.useon.com/machine/sat-t-parallel-triple-screw-extruder/* [accessed 24 March 2025]

[46] "Triple Screw Extruder", KMD Plastifizierungstechnik GmbH, *https://en.kmd-industrie.de/index.php?m=content&c=index&a=lists&catid=57* [accessed 24 March 2025]

[47] "Production Systems", ENTEX Rust & Mitschke GmbH, *https://www.entex.de/en/products/production-systems/* [accessed 24 March 2025]

[48] "MRS Extruder", Gneuß, *https://www.gneuss.com/en/polymer-technologies/extrusion/mrs-extruder/* [accessed 24 March 2025]

[49] "CONEX Invention", Conenor Ltd, *http://www.conenor.com/conex* [accessed 24 March 2025]

[50] S. Feng, A. M. Kamat, Y. Pei, "Design and fabrication of conformal cooling channels in molds: Review and progress updates", *International Journal of Heat and Mass Transfer*, 2021, vol. 171, article no. 121082, DOI: 10.1016/j.ijheatmasstransfer.2021.121082

[51] I. O. Mikulionok, "Classification of screw cooling devices of single-screw extruders for polymer materials processing (survey of designs)", *Chemical and Petroleum Engineering*, 2022, vol. 58, pp. 68–73, DOI: 10.1007/s10556-022-01057-5

[52] "ABC three-layer co-extrusion blown film machine", PLASTAR Machinery, *https://www.plastar-machine.com/en/product/ABC-Three-Layer-Co-extrusion-Blow-Film-Machine.html* [accessed 24 March 2025]

[53] "Blown film machine – PP/QPL series", Queens Machinery Co., Ltd, *https://www.queens.com.tw/en/products_i_PP_QPL_Series.html* [accessed 24 March 2025]

[54] "Plastic inspection", Dr. Schenk GmbH, *https://www.drschenk.com/products/plastic-inspection.html* [accessed 24 March 2025]

[55] "Plastic film, foil and sheets", Isra Vision, *https://www.isravision.com/en-en/industries/plastic-film-foil-sheets* [accessed 24 March 2025]

[56] "Productos para el procesamiento y producción de plásticos", Sikora, *https://sikora.net/es/categoria/plasticos/* [accessed: 24 March 2025]

5

Case Studies of Energy and Productive Improvements in Polymer Processing

Nicolás Muñoz, Omar Estrada, Iván López

The case studies presented below can be classified into two types:

- Those cases where energy consumption measurement periods last more than several days. This broad range covers both stable and unstable process conditions, allowing for energy optimization in production, quality, and process during the manufacturing of a specific product. In these cases, all necessary energy gaps can be calculated.

- Those cases where energy consumption measurements are obtained over periods of less than one hour. These short-term measurements usually cover only stable process conditions and do not provide sufficient information to determine the SEC_n. For this reason, in these cases, only the quality, process, and technology energy gaps are calculated.

5.1 Process Gap in the Injection of PE Paper Bins

5.1.1 Process Diagnosis

In a houseware injection molding company, an injection process for polyethylene (PE) bins had a stable demand of 130.2 kW and a production rate of 198.2 kg/h. The production rate was calculated based on the number of parts per cycle, the weight of the parts, and the cycle time. The product was manufactured on a 1500 t hydraulic injection molding machine with a 2-cavity hot runner mold.

The analysis of the energy consumption of the process using the EGM yielded the diagnosis shown in Figure 5.1. The image shows that the most significant gap is process + technology, which indicates that special work must be done on machine parameterization to close the process gap and isolate the technology gap. To optimize the process and isolate the technology gap, the energy consumption of each stage of the injection cycle was diagnosed. The result of this diagnosis is presented in Figure 5.2. Two-thirds of the cycle's energy consumption is associated with the dosing or plasticization phase, so special attention was paid here.

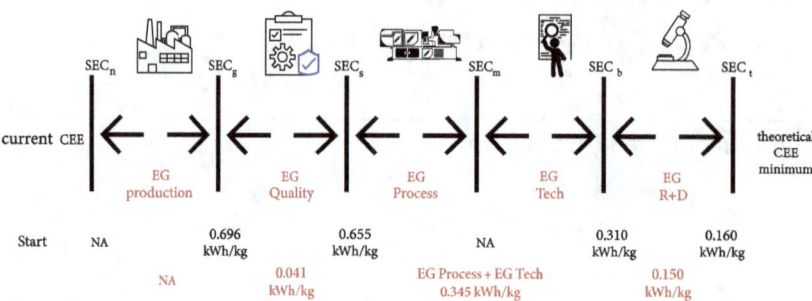

Figure 5.1 Figures were calculated to diagnose energy gaps in the PE paper bin injection process. This presents the values (in black numbers) of SEC_n, SEC_g, SEC_s, SEC_m, SEC_b, and SEC_t, as well as the respective quality, process + technology, and R&D gaps calculated, with these values in the lower central area of the SEC (in red numbers)

In this process, the cycle time is divided into stages, as described in Figure 5.3. It is observed that the product has an injection cycle where the longest stage is the dosing or plasticizing stage, which becomes the stage that limits the reduction of the cycle time. The machine is parameterized at the highest plasticizing speed allowed by the plasticizing unit, 112 rpm. This lengthens the cooling time, which is 33 s, although theoretical analysis shows that it could be 3.5 s. In other words, the part remains in the mold for 29.5 s longer than the required time due to the plasticization limitations.

The post-pressure time is longer than the theoretical cooling time and, therefore, longer than the cavity sealing time, so the post-pressure time was adjusted to optimize a few seconds of cycle time and reduce energy consumption. All energy consumed from when the injection point is sealed until the post-pressure is suspended is considered an energy expense. The engineering analysis performed using the Inyectools simulation tool allowed establishment of the theoretical cycle time for each stage, as shown in Figure 5.4.

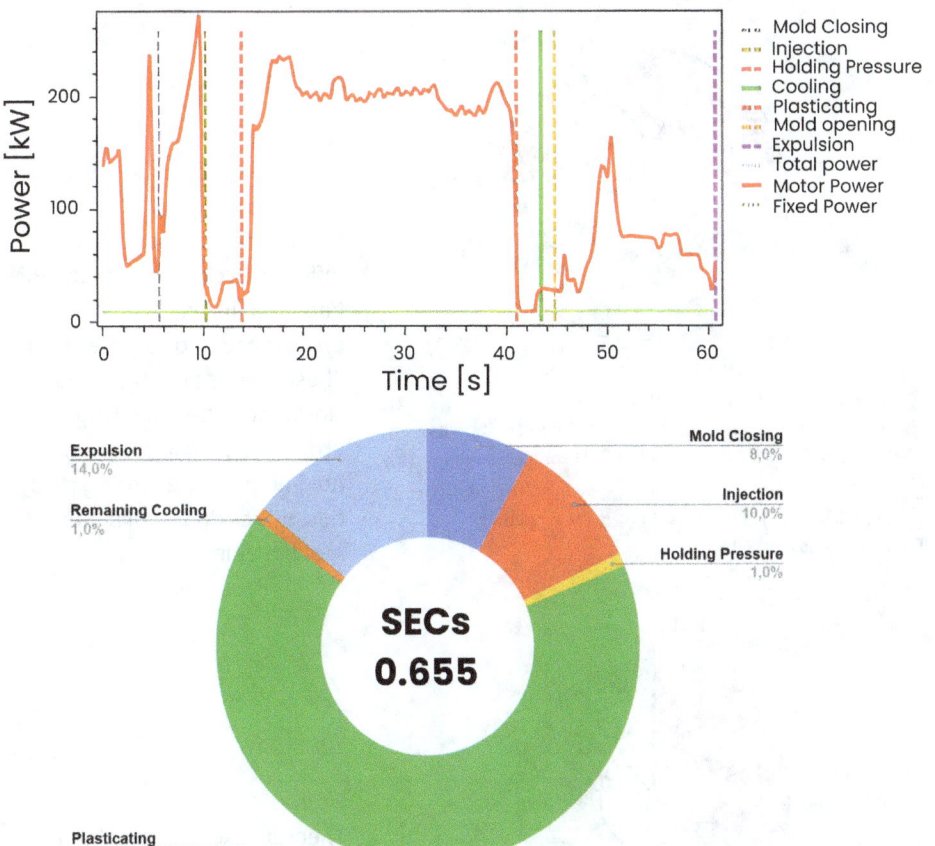

Figure 5.2 Initial specific energy consumption of the different cycle stages of the PE paper bin injection process. The upper part shows the electrical power profile (red line) during the injection cycle, with the demarcation of the end of each stage represented in a plot of power [kw] vs. time [s]. The lower part shows the specific energy consumption of the different stages of the PE paper bin injection cycle in a pie chart. These stages are organized clockwise in the following order: mold closing, injection, holding pressure, dosing delay, dosing, residual cooling, mold opening, and ejection

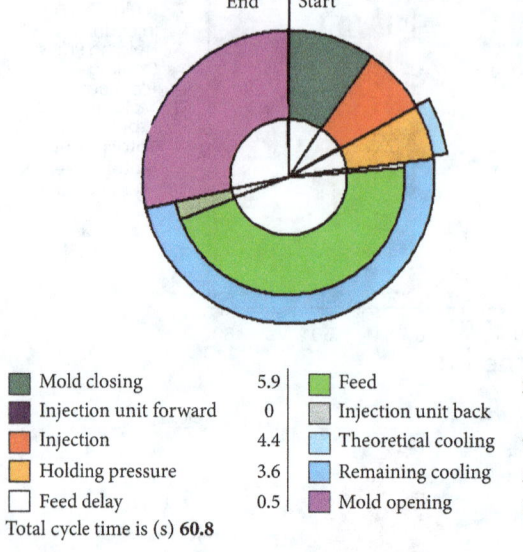

■ Mold closing	5.9	■ Feed	27.5	
■ Injection unit forward	0	▨ Injection unit back	0	
■ Injection	4.4	▥ Theoretical cooling	7.9	
■ Holding pressure	3.6	▨ Remaining cooling	3.5	
☐ Feed delay	0.5	■ Mold opening	17.4	

Total cycle time is (s) **60.8**

Figure 5.3

Initial time distribution in the different cycle stages of the PE paper bin injection molding cycle, presented as a pie chart. These stages are organized clockwise in the following order: mold closing, injection, holding pressure, dosing delay, dosing, residual cooling, and mold opening

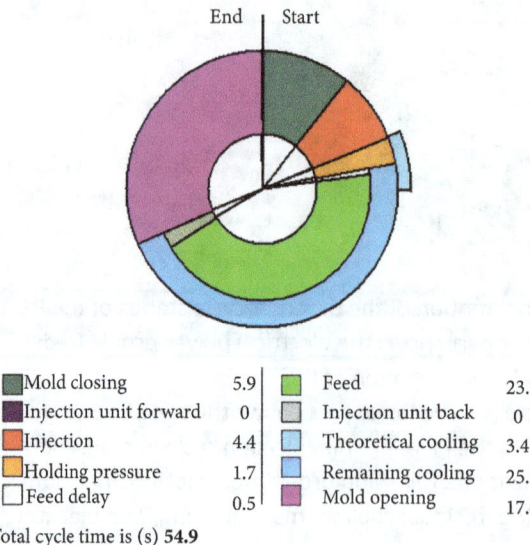

■ Mold closing	5.9	■ Feed	23.7	
■ Injection unit forward	0	▨ Injection unit back	0	
■ Injection	4.4	▥ Theoretical cooling	3.4	
■ Holding pressure	1.7	▨ Remaining cooling	25.5	
☐ Feed delay	0.5	■ Mold opening	17.4	

Total cycle time is (s) **54.9**

Figure 5.4

Theoretical cycle time distribution in the different cycle stages of the PE paper bin injection molding, presented as a pie chart. These stages are arranged clockwise in the following order: mold closing, injection, holding pressure, dosing delay, dosing, residual cooling, and mold opening

The theoretical cycle time analysis shows that the current machine capacity cannot meet the mold plasticizing speed requirements. The recommended option was therefore to move the mold to a machine with a higher plasticizing capacity. However, since there were no other machines in the company, the cycle had to be optimized with the current machine. There are two objectives: to reduce the post-pressure time and to increase the plasticizing capacity under operating conditions.

The machine was parameterized with a decreasing temperature profile (235 to 210 °C) with 3L/D of dosing and hot runner temperatures of 210 °C. This configuration shows that the output percentage of the strips presents a high demand for the initial zones. This can be solved with a flattened temperature profile that allows a better distribution of these demands by using the plasticizing work of the spindle.

5.1.2 Process Intervention

According to the findings, the intervention was decided in the plasticizing/dosing phase. For this purpose, tests were executed to characterize the energy behavior of the process, modifying only temperature profiles, plasticization back-pressure, and cooling time. The conditions were:

- Initial condition: 235 °C decreasing profile with back-pressure, and a cooling time of 29.5 s

- Condition 2: Decreasing profile of 235 °C with no back-pressure, and a cooling time of 29.5 s

- Condition 3: Decreasing profile of 235 °C with no back-pressure, and a cooling time of 25.5 s

- Condition 4: Flat profile of 225 °C with no back-pressure, and a cooling time of 29.5 s

- Condition 5: Flat profile of 225 °C with no back-pressure, and a cooling time of 25.5 s.

The SEC_s results for each of the conditions are presented in Figure 5.5. A reduction of the process energy consumption is observed for the dosing profiles without back-pressure. The SEC_s value is even lower when programming a flat temperature profile in the plasticizing unit.

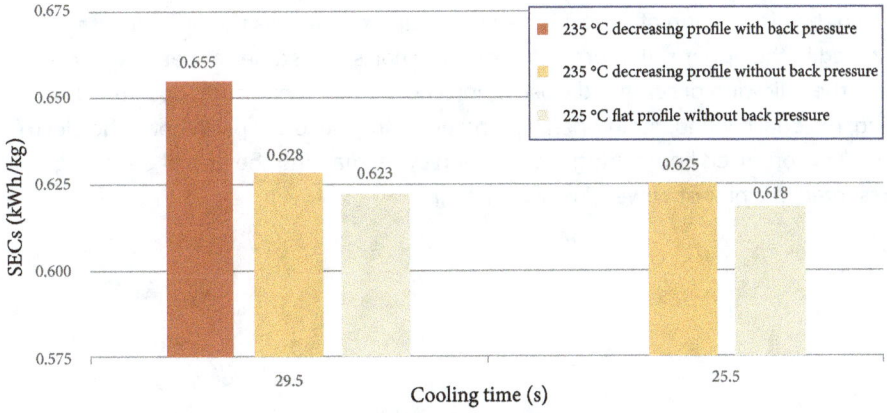

Figure 5.5 Specific energy consumption of the series of conditions evaluated in the PE paper bin injection process, presented as a bar graph. These graphs are arranged as listed above, from left to right

5.1.3 Intervention Results

The set of implemented actions allowed a reduction in SEC$_s$ from 0.655 to 0.618 kWh/kg, representing a 5.6% decrease in specific energy consumption. Likewise, the cycle time was reduced to 57 s, which meant a 6% increase in productivity. Figure 5.6 describes these results along with the new energy consumption profile.

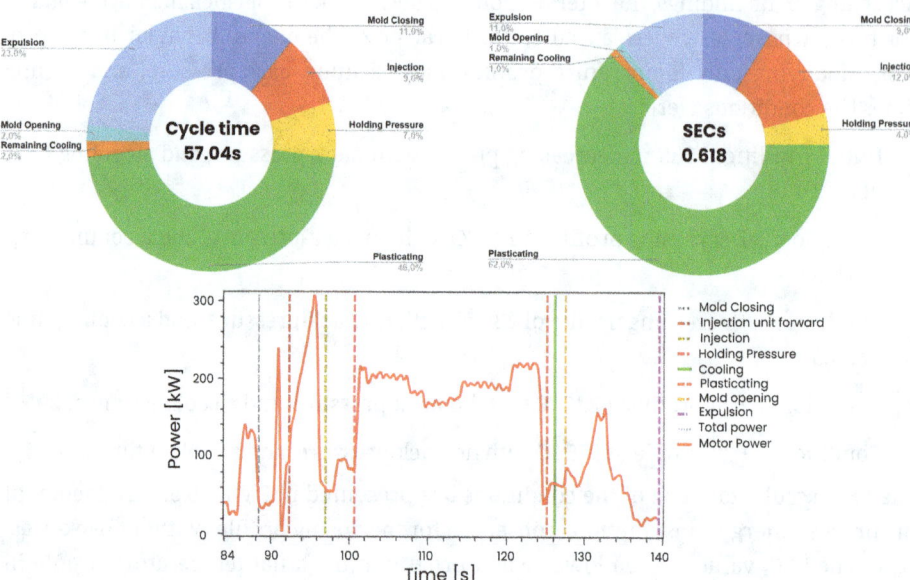

Figure 5.6 Final intervention results of the PE paper bin injection process. A pie chart showing the new cycle times is presented in the upper left corner, and a pie chart showing the distribution of specific energy consumption across the different stages is presented in the upper right corner. In both pie charts, the stages are arranged clockwise in the following order: mold closure, injection, holding pressure, dosing delay, dosing, residual cooling, mold opening, and ejection. The lower part shows the electrical power profile (red line) during the injection cycle, marking the end of each stage, represented in a plot of power [kw] vs. time [s]

The results in Figure 5.7 were obtained using the EGM to analyze the intervention. This represents a decrease in the process gap of 0.037 kWh/kg. For a production base of 90,000 kg/year, this would represent the following savings:

- Energy savings of 3300 kWh. In turn, with a value of USD 0.10/kWh, this translates into savings of USD 330/year

- A carbon footprint reduction of 0.5 t CO_2 eq/year, using 0.16438 kg CO_2 eq/kWh as the conversion factor from hydropower to CO_2 equivalent [1].

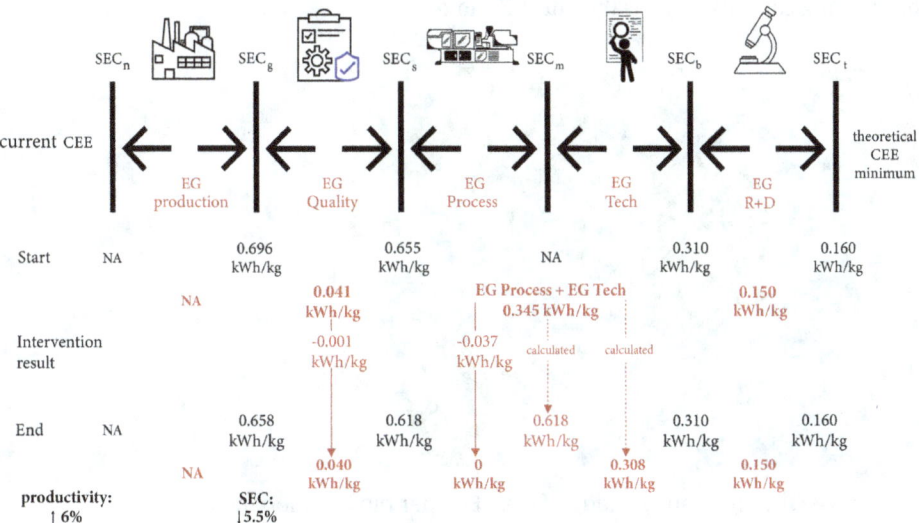

Figure 5.7 Energy gap analysis of the intervention resulting from the PE paper bin injection process. It presents the values (in black) before and after the intervention for SEC_g, SEC_s, SEC_b, and SEC_t, as well as the respective quality, process + technology, and R&D gaps calculated with these values at the two different time points in the lower central area of the SECs (in red). The results of the intervention are presented (in red) right in the middle of the two time points

Although the energy savings do not seem to generate a particularly large economic saving, the productivity improvement would allow an increase from 90,000 kg/year to 95,400 kg/year in the same production time (4,500 hours). This represents an increase of 3,200 units. If each unit had a selling cost of USD 20, this could represent an increase in annual sales of USD 64,000.

Once the intervention is finished, a very high technology gap is observed due to the condition of the plasticizing unit. To reach the optimum cycle time presented in Figure 5.4, it was recommended that the level of wear of the plasticizing unit should be checked.

5.1.4 Additional Recommendations

The mold had very low wall temperatures of 20 °C in the mold body and 30 °C in the in-
sert, as shown in the thermography presented in Figure 5.8. As this cycle is driven by
dosing, it was recommended to study the possibility of reducing mold cooling to avoid
the effects of cold cavities. A cold cavity hinders the injection process, increasing injection pressures (higher energy consumption) and the appearance of defects such as incomplete fills. In addition, reducing mold cooling reduces the cooling cycle and the cooling system's energy consumption. In the future, a new injection machine with greater
plasticizing capacity is advisable for this mold.

Figure 5.8 Thermography image of the PE paper bin injection mold

5.2 Quality and Process Gap in the PP Thermos Blow Molding Process

5.2.1 Process Diagnosis

A household product manufacturing plant had a extrusion blow-molding process for homopolymer polypropylene (PP) thermos manufacturing, as shown in Figure 5.9. This process was operated with a motor rotation speed of 1200 rpm and a screw rotation speed of
60 rpm, conditions that produced a gross mass flow (weight of the pieces together with
the burrs) of 46.9 kg/h with a cycle time of 24 s. Under these conditions, a process characterization is obtained as shown in Figure 5.10, with a SEC$_s$ of 1.369 kWh/kg.

EGM analysis of the energy consumption of the process yielded the result presented
in Figure 5.11. The graph shows that the largest gap is the technology + process gap,
with 0.749 kWh/kg, followed by the quality gap, with 0.223 kWh/kg. Special work
should therefore be done on the parameterization of the machines to close the process gap and isolate the technology gap.

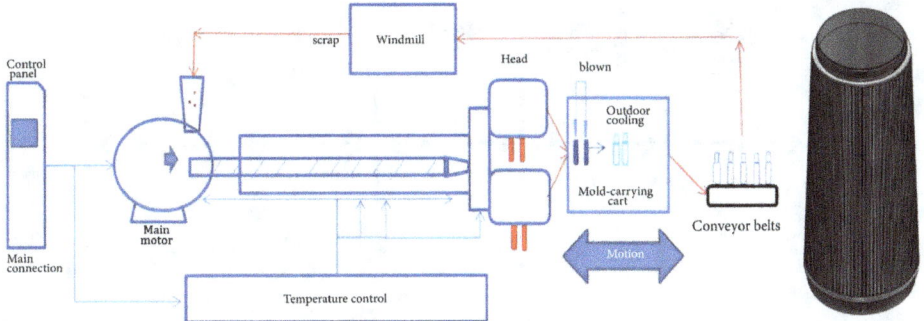

Figure 5.9 PP thermos blow molding process. The figure schematizes a double-die, double-station line with two two-cavity molds at each station. One movement cycle of the two stations generates the production of 4 bottles

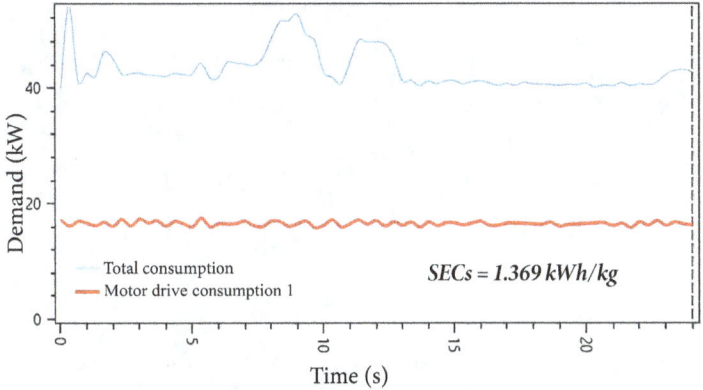

Figure 5.10 Initial SEC$_s$ of the PP thermos extrusion blow molding process, calculated with stable power. This is presented in a plot of power vs. time, with the total power shown as a blue line and the motor power as a red line

Variations in the behavior of the totalizer and the motor were found after characterizing the power demand during several cycles. This is observed in Figure 5.12, where the significant variation of cycles total demand in different cycles is evidenced. The behavior of a stable process should present stable demand profiles. This may be due to reasons such as:

- Plasticizing unit wear and tear

- Problems in the operation of the motor or its components

- Malfunctioning of the parison programmer

- Variation in the quality of the raw material due to the incorporation of the regrind material

- Operating condition of low motor rotation speed.

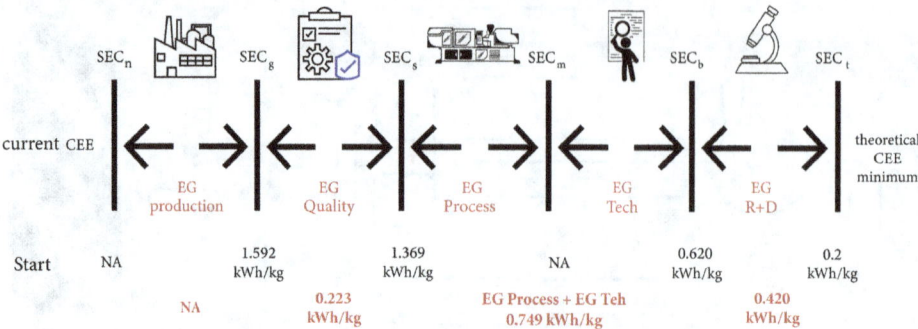

Figure 5.11 Initial diagnosis of the energy gaps in the PP thermos extrusion blow-molding process. This section presents the values (in black) for SEC_g, SEC_s, SEC_n, SEC_b, and SEC_t, and the respective quality, process + technology, and R&D gaps calculated with these values in the lower central SEC area (in red)

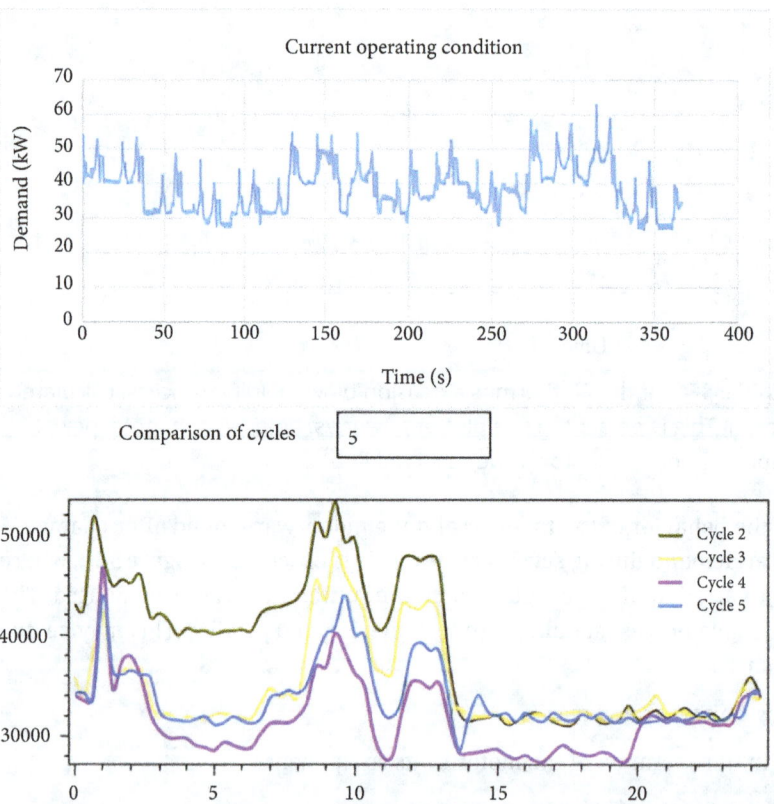

Figure 5.12 Initial comparison of electrical power demand stability for four initial cycles of the PP thermos extrusion blow-molding process. The top section shows the totalized electrical power profile (blue line) during the sampled period in a power [kW] vs. time [s] diagram. The bottom section shows the power comparison in a cycle power [kW] vs. time [s] diagram during four overlapping cycles

5.2.2 Process Intervention

Among the hypotheses made about the variable behavior of the motor's energy demand, the leading cause was thought to be the low spindle rotation speed. Tests were carried out by increasing this parameter, taking the spindle to rotate at 63.5 rpm (equivalent to going from 1200 rpm to 1270 rpm in the motor). This reduced the cycle time from 24 s to 22 s. This intervention allows for obtaining a more stable behavior of total cycle demand, as shown in Figure 5.13.

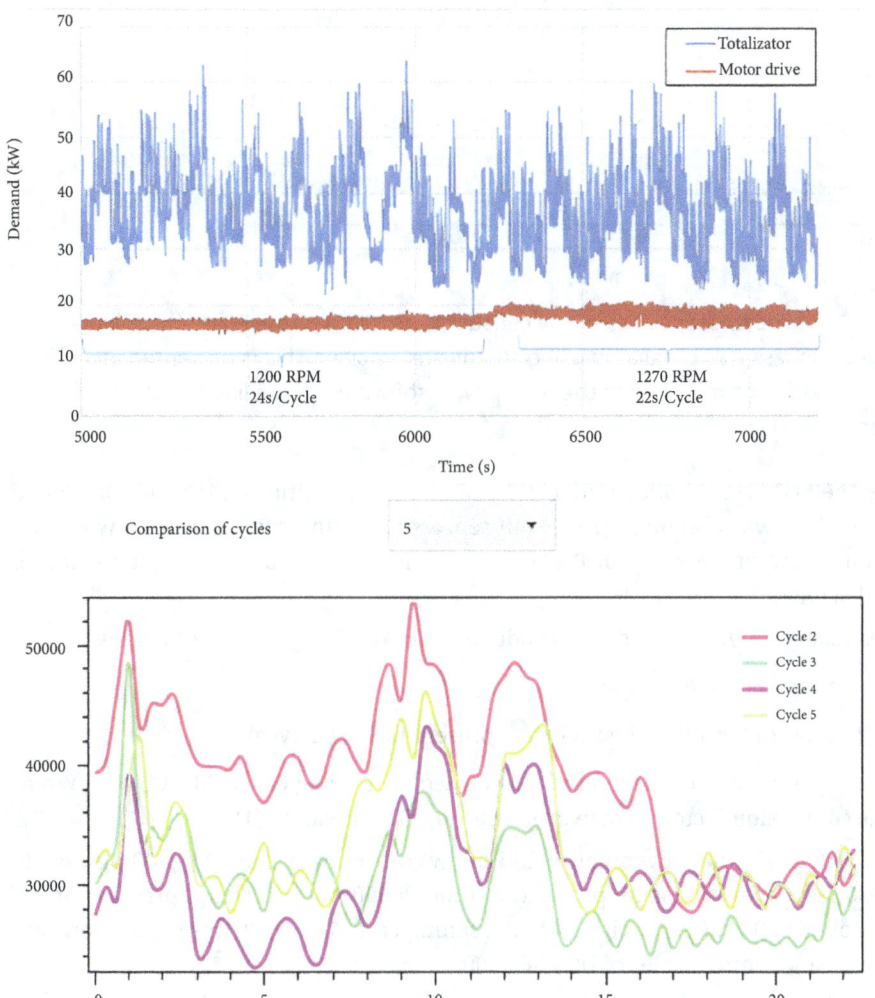

Figure 5.13 A final comparison of the stability of electrical power for four initial cycles of the PP thermos extrusion blow-molding process. The top section shows the totalized electrical power profile (blue line) during the sampled period in a plot of power [kW] vs. time [s]. The bottom section shows the power comparison in a plot of cycle power [kW] vs. time [s] during four overlapping cycles

5.2.3 Intervention Results

With the modifications made to the parameters, an SEC_s of 1.085 kWh/kg was obtained, as shown in Figure 5.14. This is 20.8% lower than the initial value, with an increase in production of 8.3% due to the reduction in cycle time.

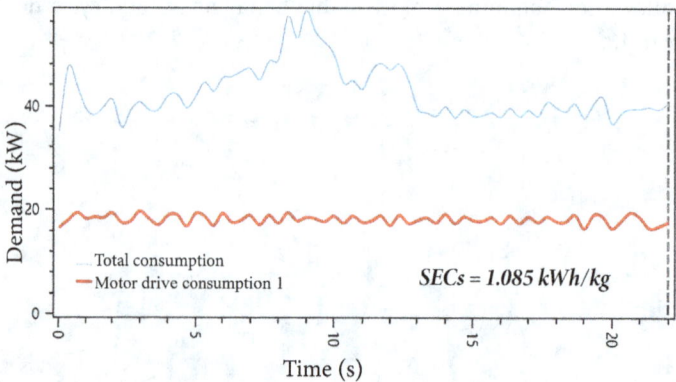

Figure 5.14 The final specific energy consumption (SEC_s) of the PP thermos blowmolding process was calculated using steady-state power. This is presented in a plot of power [kW] vs. time [s], with the total power shown as a blue line and the motor power as a red line

When the EGM was applied under the new process conditions, the result presented in Figure 5.15 was obtained. This result represents an intervened process with a reduction of the process gap of 0.284 kWh/kg and a decrease in the quality gap of 0.047 kWh/kg.

The estimated savings, based on a production of 110,000 kg/year, are as follows:

- Energy savings of 36,410 kWh

- For USD 0.10 per kWh, the savings represent USD 3,641/year

- A carbon footprint reduction of 5.3 t CO_2 eq/year, using 0.16438 kg CO_2 eq/kWh as the conversion factor from hydropower to CO_2 equivalent [1].

This productivity improvement would allow an increase from 110,000 kg/year to 119,130 kg/year in the same production time (4,000 hours). This represents an increase of 96,000 units. If each unit had a selling cost of USD 10, this could represent an increase in annual sales of USD 960,000.

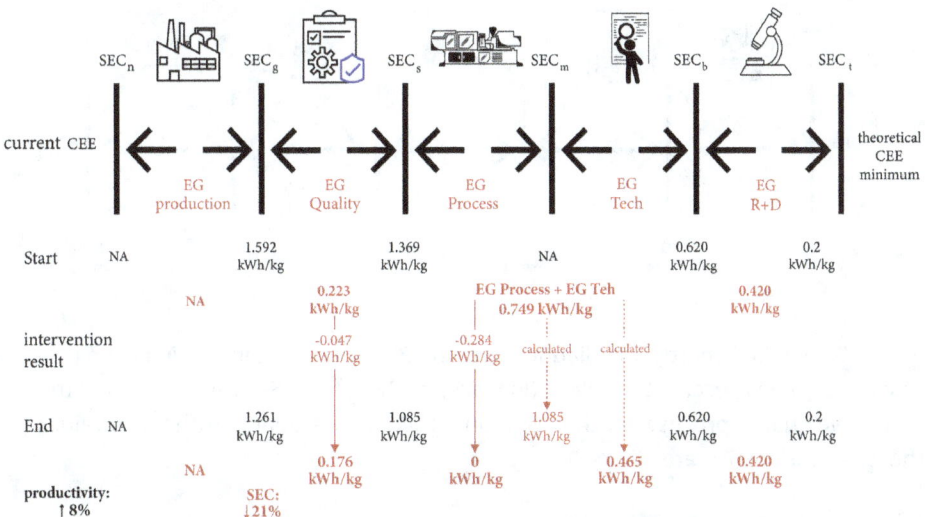

Figure 5.15 Energy gap analysis of the PP thermos extrusion blow-molding process intervention results. This section presents the values (in black) before and after the intervention for SEC_g, SEC_s, SEC_b, and SEC_t, as well as the respective quality, process + technology, and R&D gaps calculated with these values at the two different time points in the lower central area of the SEC_s (in red). The results of the intervention are presented (in red) right in the middle of the two time points

5.3 Process Gap in Injection of PE Applicator Cannula

5.3.1 Process Diagnosis

A medical products company was carrying out an injection process of a 2.9 g polyethylene (PE) product. The material was processed in an injection molding machine of 1800 kN and 45 mm screw diameter, with a 16-cavity mold with a mixed hot and cold runner system.

The analysis of the energy consumption process using the EGM shows the diagnosis presented in Figure 5.16, which shows that the most significant gap is the process + technology gap. Therefore, special work needed to be done on the parameterization of the machines to close the process gap and isolate the technology gap. Reviewing the machine's operation, the parameters presented in Table 5.1 were observed.

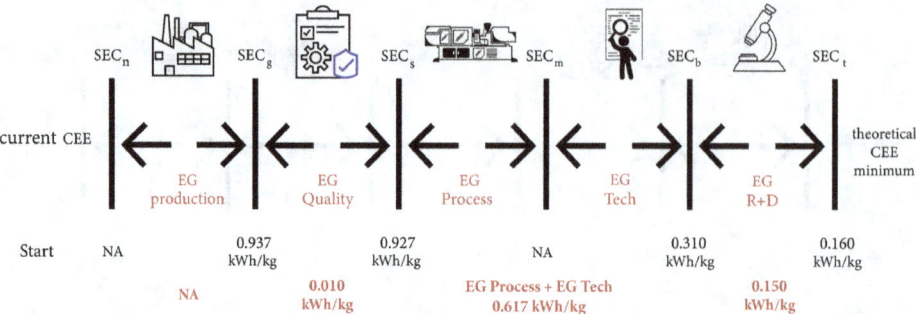

Figure 5.16 Initial energy gap diagnosis for the PE applicator cannula injection process. This graph presents the values (in black) for SEC$_g$, SEC$_s$, SEC$_b$, and SEC$_t$, and the respective quality, process + technology, and R&D gaps calculated with these values in the lower central SEC area (in red)

Table 5.1 PE Applicator Nozzle Injection Process Conditions

Parameter	Value
Injection speed profile [%]	94/94/94
Injection position profile [mm]	18/17
Injection pressure profile [bar]	113
Switching point [mm]	8
Post-pressure profile [bar]	110
Post-pressure time [s]	0.9
Programmed cooling time [s]	3
Temperature profile [°C]	220/225/215/180
Plasticizing speed [%]	99
Back-pressure [bar]	0
Dosing [mm]	67
Decompression travel [mm]	10
Mass cushion [mm]	18.5
Final position [mm]	77
Plasticizing L/D [—]	1.49
Hot runner temperature [°C	279/278/248/302/299/299
Mold temperature [°C]	—
Cycle time [s]	11.7

During the on-machine analysis, it was found that the control pressure of 113 bar limited the machine's operation for volumetric filling without justification. In addition, given the low percentage of spindle occupancy (1.5*L/D*), it was considered that the temperature profile could be more gradual and increase in its totality. For its part, the temperature control of the hot runner system operated at very high temperatures, mostly with values above 280 °C. The function of the temperature control in these systems is to maintain the plasticizing unit nozzle temperature (220 °C). Higher temperatures can lead to material degradation and increase the energy consumption of the process. These process conditions resulted in an injection cycle time of 11.7 s, distributed as shown in Figure 5.17.

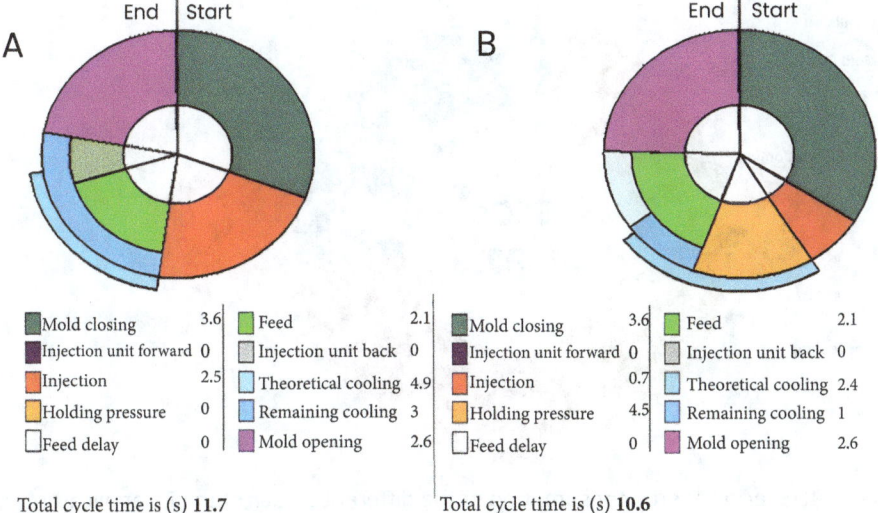

A			
■ Mold closing	3.6	■ Feed	2.1
■ Injection unit forward	0	■ Injection unit back	0
■ Injection	2.5	■ Theoretical cooling	4.9
■ Holding pressure	0	■ Remaining cooling	3
☐ Feed delay	0	■ Mold opening	2.6

Total cycle time is (s) **11.7**

B			
■ Mold closing	3.6	■ Feed	2.1
■ Injection unit forward	0	■ Injection unit back	0
■ Injection	0.7	■ Theoretical cooling	2.4
■ Holding pressure	4.5	■ Remaining cooling	1
☐ Feed delay	0	■ Mold opening	2.6

Total cycle time is (s) **10.6**

Figure 5.17 Time distribution in the different stages of the injection cycle for the PE applicator cannula, with the initial conditions on the left (a) and optimized by simulation in Injectools software on the right (b). These stages are organized clockwise in the following order: mold closing, injection, holding pressure (except for a, which does not have this stage), dosing delay, dosing, remaining cooling, and mold opening

The result presented in Figure 5.18 was obtained by studying the demand and obtaining the cycle's energy footprint. The injection and dosing phases account for more than 50% of the total consumption, so special emphasis was placed on these stages for the cycle optimization analysis.

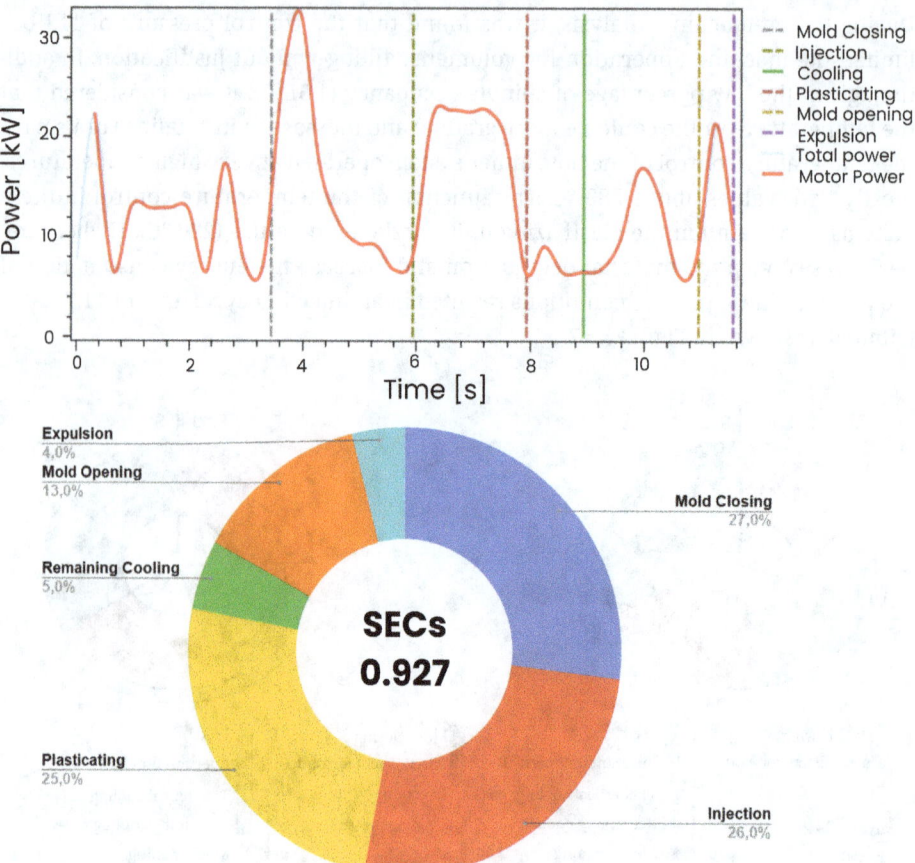

Figure 5.18 Specific energy consumption of the different stages of the PE applicator cannula injection cycle. The upper section presents the electrical power profile (red line) in a plot of power [kW] vs. time [s] during the injection cycle, with the demarcation of the end of each stage. The lower section presents the specific energy consumption of the different stages of the PE applicator cannula injection cycle in a pie chart. These stages are organized clockwise in the following order: mold closure, injection, holding pressure, dosing delay, dosing, residual cooling, mold opening, and ejection

5.3.2 Process Intervention

Tests were carried out to study the energetic behavior of the process by modifying different operating parameters. The test conditions are described below:

- Initial condition: No parameter changes

- Condition 2: Injection setting with injection control pressure of 180 bar

- Condition 3: Injection setting and cylinder temperature profile of 220/210/200/180 °C

- Condition 4: Injection set-up, cylinder temperature profile, and hot runner temperatures unified at 240 °C.

Figure 5.19 presents the specific energy consumption evaluation results under these conditions. These results show an appreciable reduction in process energy consumption with the injection control pressure setting of Condition 2. Consumption continues to decrease, although to a lesser extent, with the setting of the cylinder temperature profile and the unified hot runner temperature setting at 240 °C.

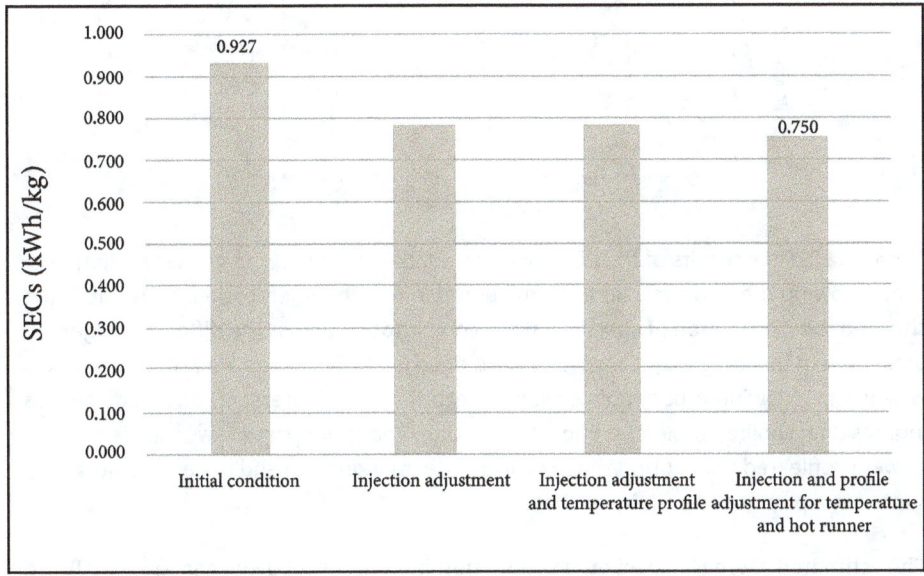

Figure 5.19 Specific energy consumption for the series of conditions evaluated during PE injection. This bar chart is arranged as listed above, from left to right

5.3.3 Intervention Results

The intervention resulted in a change of the SEC from 0.927 to 0.75 kWh/kg, which means a decrease in the specific energy consumption of 19.1%. Likewise, the cycle time decreased by up to 11 s, representing an increase in productivity of 6.4%. Figure 5.20 describes these results along with the new energy consumption profile. Figure 5.21 presents the result of the energy gap analysis for the process after the intervention. This represents a decrease in the process gap of 0.177 kWh/kg.

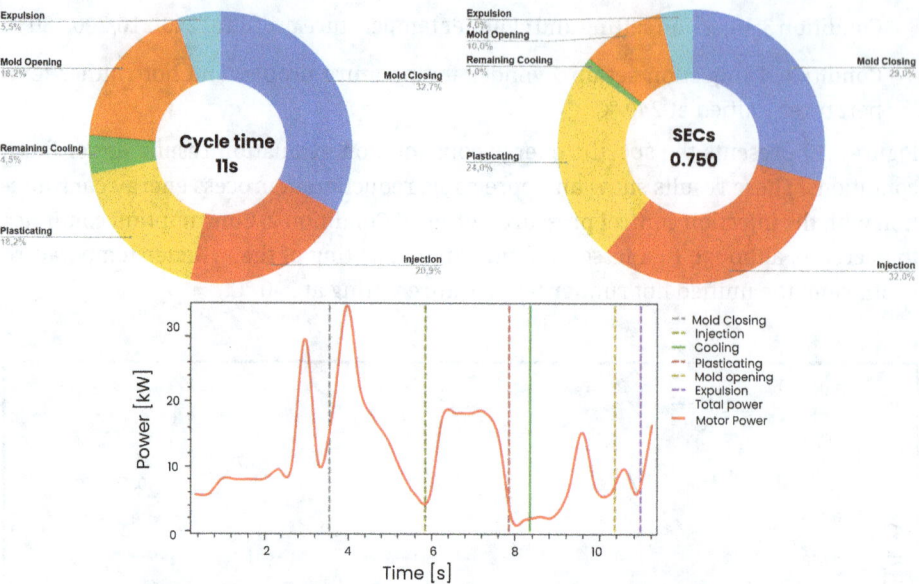

Figure 5.20 Final results of the PE applicator nozzle injection process intervention. A pie chart showing the new cycle times is presented in the upper left corner, and a pie chart showing the distribution of specific energy consumption across the different stages is presented in the upper right corner. In both pie charts, the stages are arranged clockwise in the following order: mold closure, injection, holding pressure, dosing delay, dosing, residual cooling, mold opening, and ejection. The lower part shows the electrical power profile (red line) during the injection cycle, marking the end of each stage, represented in a plot of power [kW] vs. time [s]

The estimated savings, based on a production base of 100,000 kg/year, are as follows:

- Energy savings of 17,900 kWh

- For USD 0.10/kWh, the savings represent USD 1,790/year.

A carbon footprint reduction of 2.6 t CO_2 eq/year, using 0.16438 kg CO_2 eq/kWh as the conversion factor from hydropower to CO_2 equivalent [1].

This productivity improvement would allow an increase from 100,000 kg/year to 106,400 kg/year in the same production time (7,200 hours). This represents an increase of 2,210,000 units. If each unit had a selling cost of USD 0.05, this could represent an increase in annual sales of USD 110,500.

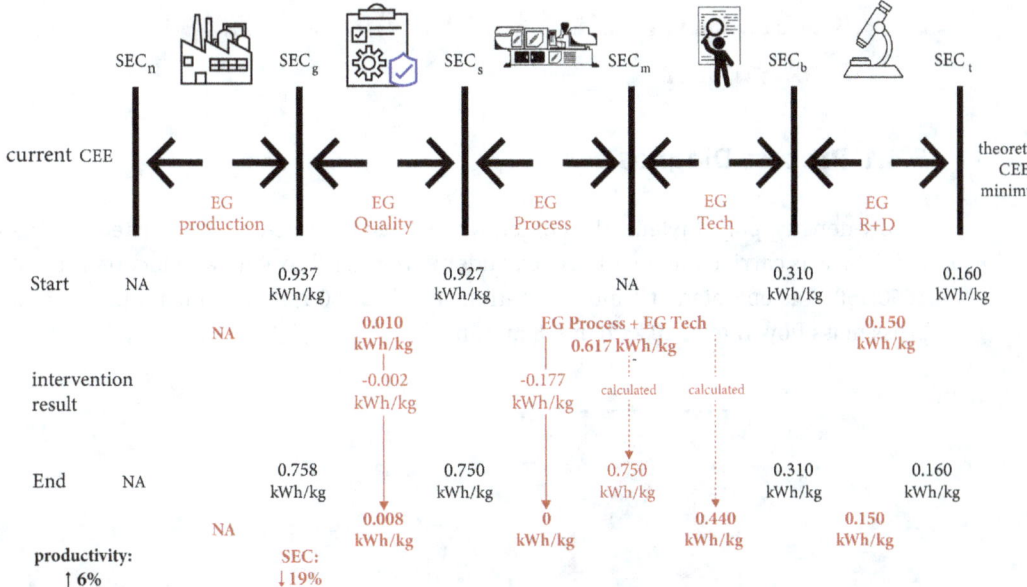

Figure 5.21 Energy gap analysis of the outcome of the PE cannula injection intervention. This section presents the values (in black) before and after the intervention for the SEC_g, SEC_s, SEC_b, and SEC_t, as well as the respective quality, process + technology, and R&D gaps calculated with these values at the two different time points in the lower central area of the SECs (in red). The results of the intervention are presented (in red) right in the middle of the two time points

5.3.4 Additional Recommendations

A cooling time greater than three times the dosing time was observed, which made the process less than optimal. A short post-pressure time with respect to volumetric filling was also observed, which resulted in a process in which practically no gravimetric filling is performed. This represents a problem for the efficiency of the process and the quality of the parts.

5.4 Process Gap in LDPE Blown Film with Zipper Coextrusion

5.4.1 Process Diagnosis

A low-density polyethylene (LDPE) blown extrusion process, as depicted in Figure 5.22, was carried out on a 15 HP extruder with a gearbox with a reduction ratio of 1650:150. It is operated at a motor rotation speed of 500 rpm, at which it produces a gross mass flow rate of 10 kg/h or 6.8 m/min.

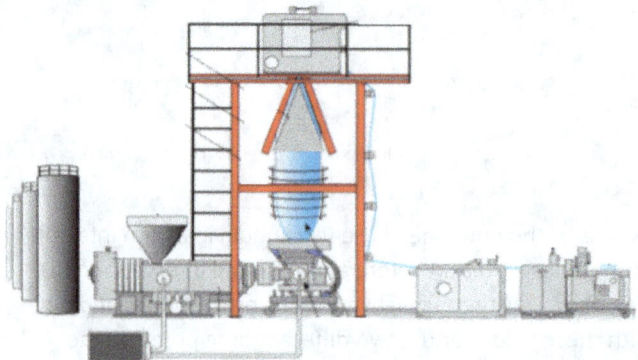

Figure 5.22 LDPE blown film with zipper forming by coextrusion

Figure 5.23 presents the result of the energy consumption analysis of the process using the EGM. The largest gap is the technology + process gap, with a value of 0.631 kWh/kg. Special work needed to be done on the parameterization of the machines to close the process gap and isolate the technology gap.

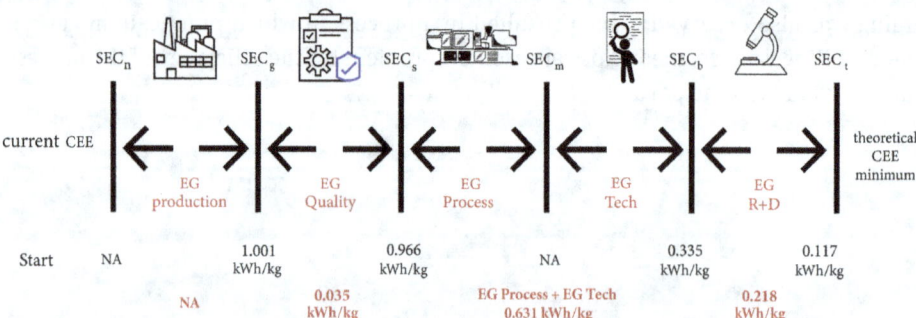

Figure 5.23 Initial diagnosis of the energy gaps in the LDPE zipper co-extrusion process. This section presents the values (in black) for SEC_g, SEC_s, SEC_b, and SEC_t, and the respective quality, process + technology, and R&D gaps calculated with these values in the lower central SEC area (in red)

5.4.2 Process Intervention

Given the characteristics of the motor and machine, it was considered that the machine was being underutilized. The motor offers a speed of 1,760 rpm, with a gearbox ratio of 1650:150. Because of this, frequency variation tests were run, during which the demand and mass flow were characterized with respect to the motor frequency. As can be seen in Figure 5.24, motor rotation speeds were set from 500 rpm to 750 rpm with 50 rpm increments.

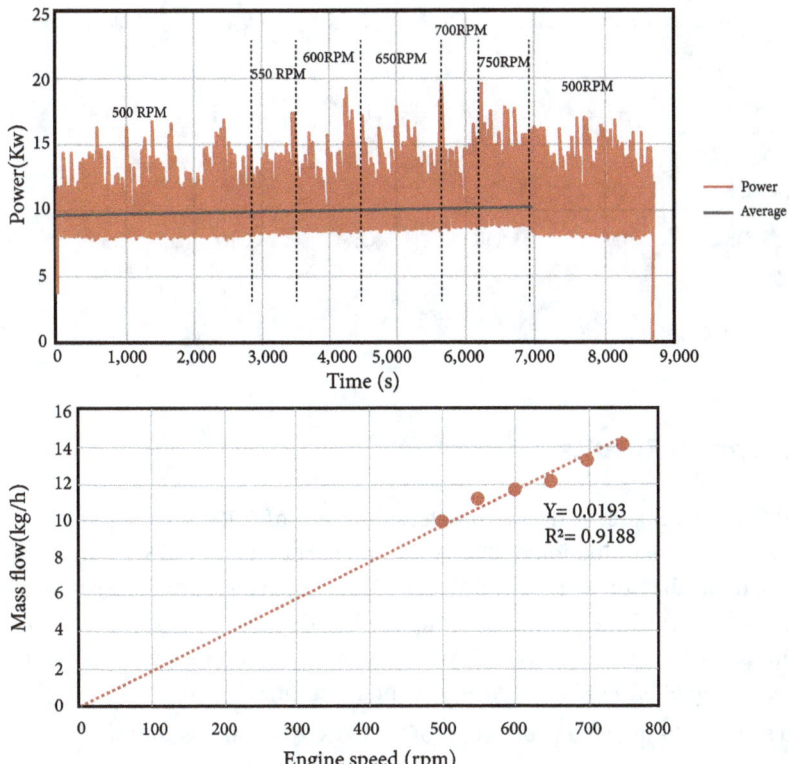

Figure 5.24 Frequency variation test of the LDPE zipper co-extrusion line motor. The top section shows the total (red line) and average (gray line) electrical power profiles during motor speed changes in a plot of time [s] vs. power [kW]. The bottom section shows the variation in mass flow rate with respect to motor speed (dots and red trend line) in a plot of motor speed [rpm] vs. mass flow rate [kg/h]

Data on the main motor speed, temperature profile, and pulling speed were recorded. Film thickness was also measured for each speed, and the mass flow rate was estimated for each operating condition. These results are presented in Table 5.2.

Table 5.2 Results of Motor Frequency Variation of the LDPE Zipper Co-extrusion Line

Parameter	Condition 1	Condition 2	Condition 3	Condition 4	Condition 5	Condition 6
Motor speed [rpm]	500	550	600	650	700	750
Current [A]	33.8	35.8	36	36.6	37	37.4
Discharge pressure [kPa]	1,355	1,365	1,375	1,400	1,405	1,425
Volumetric flow [m^3/min]	6.8	7.2	8.1	8.8	10	10.7
Mass flow [kg/h]	10	11.2	11.8	12.1	13.3	14.1
Zipper	OK	OK	OK	OK	OK	OK
Thickness	OK	OK	OK	OK	OK	OK
Special conditions	—	—	Increased airflow	—	—	—

5.4.3 Intervention Results

With the modifications made to the parameters, an SEC_s of 0.722 kWh/kg was obtained, which represents a reduction of 25.3% with respect to the initial SEC_s. In addition, an increase in production of 41% was achieved due to the increased mass flow of the intervened reference. When the EGM is applied under the new process conditions, the result presented in Figure 5.25 was obtained. The analysis shows a reduction of the process gap of 0.244 kWh/kg after the intervention.

The estimated savings, based on a production of 70,000 kg/year, are as follows:

- Energy savings of 17,640 kWh

- For USD 0.10/kWh, the savings represent USD 1,764/year

- A carbon footprint reduction of 2.6 t CO_2 eq/year, using 0.16438 kg CO_2 eq/kWh as the conversion factor from hydropower to CO_2 equivalent [1].

This productivity improvement would allow an increase from 70,000 kg/year to 98,700 kg/year in the same production time. This represents an increase of 28,700 kg. If each kg generates profits of USD 0.12, this represents an increase in annual profits of USD 3,444.

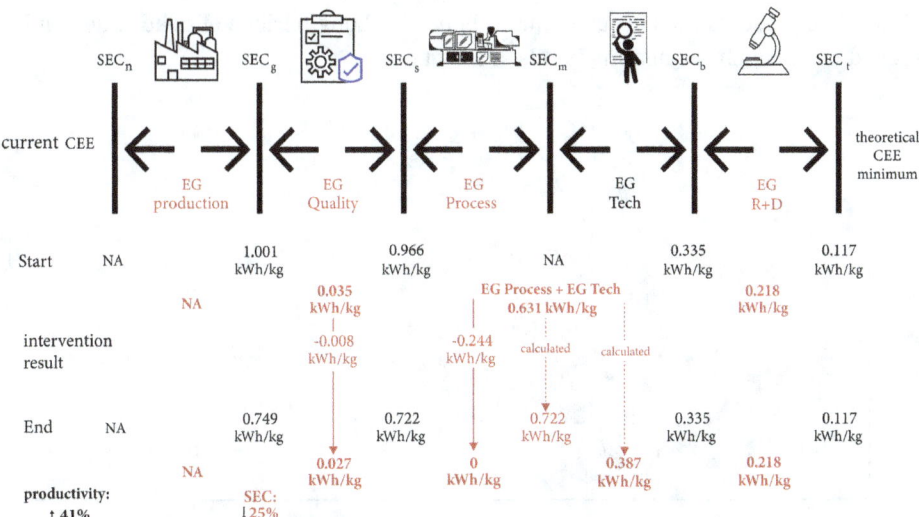

Figure 5.25 Energy gap analysis of the results of the intervention in the LDPE zipper co-extrusion process. The values (in black) before and after the intervention are presented for SEC_g, SEC_s, SEC_n, SEC_b, and SEC_t, as well as the respective quality, process + technology, and R&D gaps calculated with these values at the two different time points in the lower central area of the SEC_s (in red). The results of the intervention are presented (in red) right in the middle of the two time points

5.4.4 Additional Recommendations

Although increasing the motor frequency further was impossible due to process stability and product quality defects, this aspect could have been analyzed by incorporating a curved ring around the bubble. Therefore, the company was advised to implement such a ring to reduce oscillations, expand the cooling area, and maintain the quality of the extruded product.

5.5 Production and Process Gap in a Plastic Moulded Profile Process

5.5.1 Process Diagnosis

The profile molding process uses a single-screw extruder to fill a mold that has the shape of the profile being manufactured. By this method, the profiles are not hollow, and the process is discontinuous. The extruder stops while the mold is being changed.

Figure 5.26 presents the energy demand behavior. Long periods of dead time can be seen, during which the heating bands are kept on.

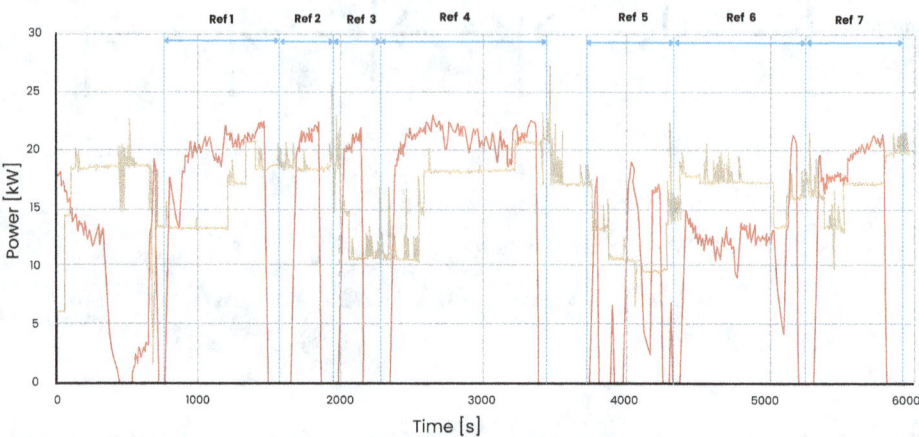

Figure 5.26 Power demand behavior of the extruder during the production of the different references of plastic wood. This is presented in a plot of power [kW] vs. time [s], with the motor power shown as a red line and the heating bands power as an orange line

Figure 5.27 presents the result of analyzing the energy consumption process using the EGM. The highest gap is the production gap.

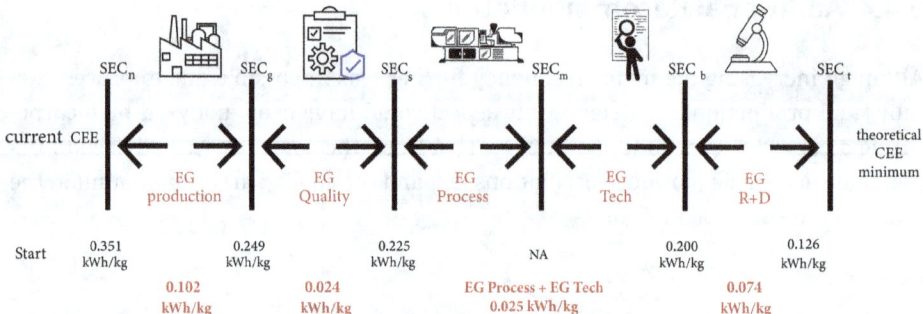

Figure 5.27 Initial diagnosis of the energy gaps in plastic wood/profile extrusion process. This section presents the values (in black) for SEC_n, SEC_g, SEC_s, SEC_b, and SEC_t, and the respective quality, process + technology, and R&D gaps calculated with these values in the lower central SEC area (in red)

During the analysis of production downtime, which represented approximately 45% of total operating time, the main cause was mold changeover. This problem was related to the die system, which required dismantling the full mold before installing the next one. This process creates significant delays, as the die must be air-cooled before the operator can safely handle it.

The next most significant energy gap identified in the analysis is the process + technology gap. This gap may be related to using recycled material, which could have accelerated the wear of the screw and plasticizing unit and the energy consumption of the heaters during idle times.

5.5.2 First Intervention

To attack the production energy gap, new technology was implemented in the production process. Specifically, a die was incorporated to connect two molds, as shown in Figure 5.28. This die has a pressure sensor that controls the feed flow through one outlet at a time so that once one mold is filled, the other mold connected to the other outlet of the die starts to be filled. Thanks to this, the cooling required for mold removal does not interrupt production and makes the process more continuous.

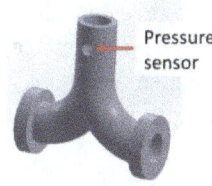

Pressure sensor

Figure 5.28
Die for the connection of two molds, including the pressure sensor

5.5.3 Second Intervention

Similarly, the process gap was closed by implementing a technological improvement, namely replacing the on/off heating system with a cartridge system. Additionally, thermal insulation of the die and plasticizing unit was carried out to optimize energy efficiency and reduce heat losses.

However, it was observed that the extruder's performance did not meet operational expectations. During the production of the parts analyzed, it was determined that most of the energy (55%) was consumed by the heating bands suggesting a high degree of wear on the screw. This wear prevents efficient plasticization of the material, a stage that depends on the heating system.

The effects of wear were also evident in obtaining the extrusion lines, visible in Figure 5.29, and in the low correlation obtained in the fit coefficient analysis. Furthermore, when evaluating the power transmission ratio of the extruder as a function of its diameter, it was found that the extrusion capacity of the equipment corresponded to only 50% of the expected mass flow. These findings confirm the deterioration of the plasticizing unit impacting the efficiency of the process.

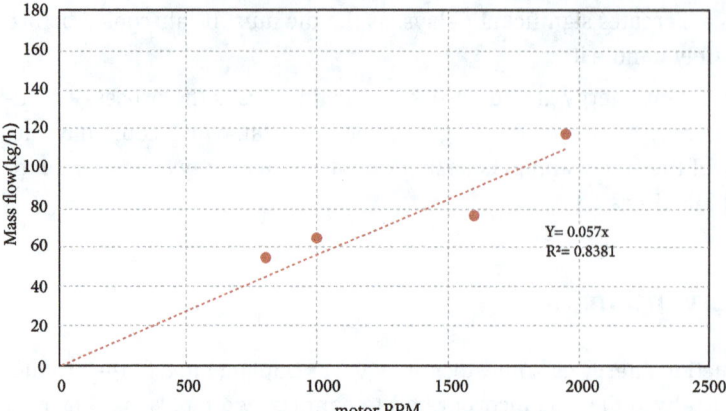

Figure 5.29 Extrusion lines operating on the equipment. The variation in mass flow rate with respect to motor speed (dots and red trend line) is presented in a plot of motor speed [rpm] vs. mass flow rate [kg/h]

The analysis of the behavior of the total electrical demand and the different components of the extruder at different mass flows, as shown in Figure 5.30, revealed a low correlation coefficient for the extruder. When specifically evaluating the performance of the motor and the resistors, it was observed that the latter presented a similarly low correlation. This fact suggests that the irregular energy consumption in the extruder is directly related to the operation of the heating belts.

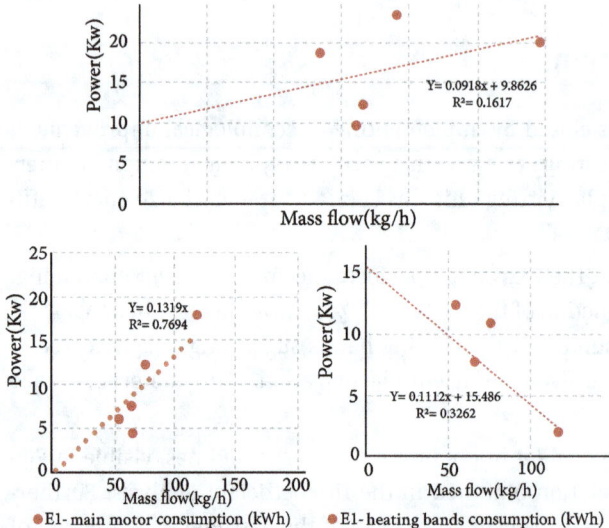

Figure 5.30 The behavior of the total electrical demand and extruder components vs. mass flow rate. This is represented at the top for the line, at the bottom left for the motor, and at the bottom right for the heating elements. The power variation (dots and red trend line) is presented in a plot of power [kW] vs. mass flow rate [kg/h]

5.5.4 Intervention Results

Figure 5.31 presents the result of applying the EGM after the process interventions. In this production process, it was possible to close the production gap by 0.053 kWh/kg and the process gap by 0.014 kWh/kg. These modifications represented an estimated average production increase of 23% and an estimated specific energy consumption reduction of up to 20%.

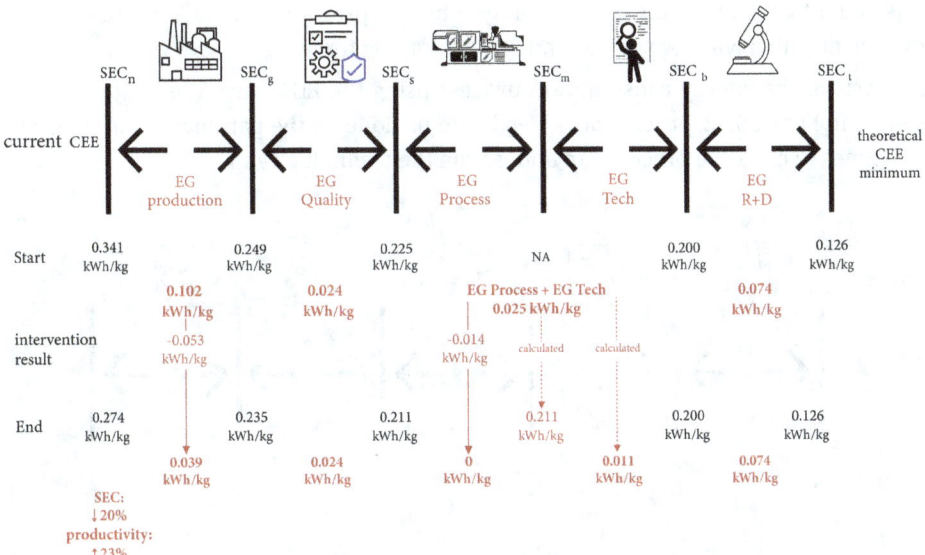

Figure 5.31 Energy gap analysis of the result of the intervention in the extrusion process of plastic wood/profiles. This section presents the values (in black) before and after the intervention for SEC_n, SEC_g, SEC_s, SEC_b, and SEC_t, as well as the respective quality, process + technology, and R&D gaps calculated with these values at the two different time points in the lower central area of the SECs (in red). The results of the intervention are presented (in red) right in the middle of the two time points

The estimated savings, based on a production of 220,000 kg/year, are as follows:

- Energy savings of 14,740 kWh

- For USD 0.10/kWh, the savings represent USD 1,474/year

- A carbon footprint reduction of 2.2 t CO_2 eq/year, using 0.16438 kg CO_2 eq/kWh as the conversion factor from hydropower to CO_2 equivalent [1].

This productivity improvement would allow an increase from 220,000 kg/year to 270,600 kg/year in the same production time. This represents an increase of 70,600 kg. If each kg generates profits of USD 0.25, this could represent an increase in annual profits of USD 17,650.

5.6 Quality and Process Gap in Injection of Automotive Parts

5.6.1 Process Diagnosis

A company was producing plastic parts for automobiles. One of the references produced is a 1.3 kg part made of polypropylene (PP) reinforced with fiberglass, which is injected into a mold with a hot runner system. The product is manufactured on a hydraulic machine with two drives and 700 t clamping force.

Analysis of the energy consumption process using the EGM gave the diagnosis presented in Figure 5.32. Special work needed to be done on the parameterization of the machines to close the process gap and isolate the technology gap.

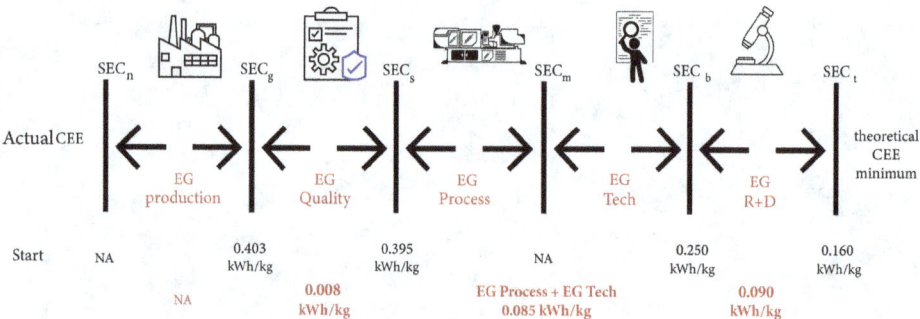

Figure 5.32 Initial diagnosis of energy gaps in automotive part production. This table presents the values (in black) for SEC_g, SEC_s, SEC_b, and SEC_t, and the respective quality, process + technology, and R&D gaps calculated with these values in the lower central SEC area (in red)

The product was being manufactured with the parameters presented in Table 5.3. The machine has two drives, which have different functions during the cycle:

- Drive 1 oversees the screw movements during the filling, packaging pressure, and plasticizing stages

- Drive 2 has mold movements as its primary function, but it also supports drive 1 during the filling stage.

For the study, the energy demands of drive 1, drive 2, and the whole machine (totalizer) were measured. The energy demand of the heating band can be estimated by subtracting the drive demands from the totalizer demand. Thus, the energy power profile of the cycle presented in in Figure 5.33 was obtained.

Table 5.3 Injection Molding Machine Operating Parameters

Parameter	Initial condition
Cooling time [s]	33
Barrel temperature [°C]	210
Hot runner temperature [°C]	232
Plasticizing time [s]	20
Plasticizing speed [%]	80
Filling speed profile [%]	39.2/48
Packing pressure profile [kg/m^3]	65/55
Packing time [s]	1.4/1.3
Cycle time [s]	91.5

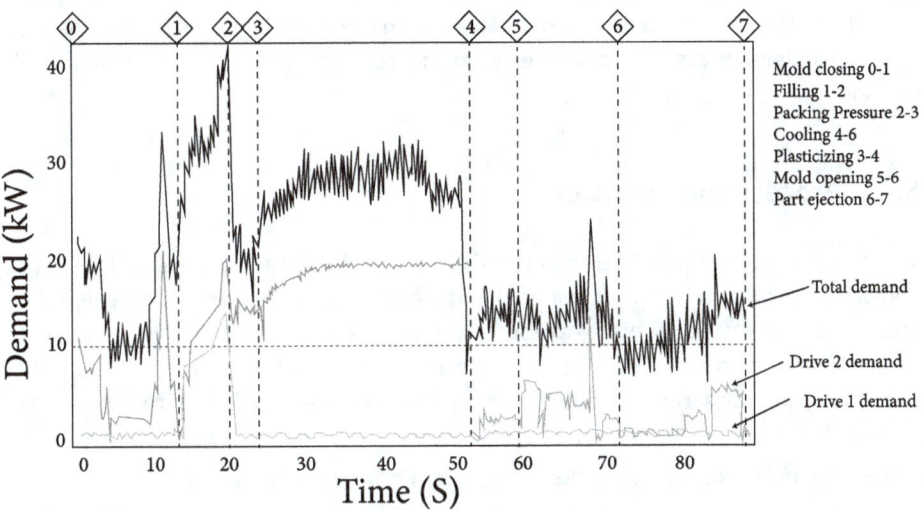

Figure 5.33 Energy power profile of the injection cycle of an automotive part. The profile of total electrical power (black line) and of the two motors (respectively marked gray lines) during the injection cycle are presented, with the demarcation of the end of each stage at the top represented in a plot of time [s] vs. power [kW]

Figure 5.34 shows the distribution of the consumption at each stage of the injection cycle. Although the plasticizing stage has the highest energy consumption, its consumption is within the common range for long parts. Usually, the plasticizing stage consumes 35–45% of the total energy consumption of the cycle, and the remaining cooling can represent 20–30%. The next stage in terms of consumption is the mold filling stage, i.e., injection and holding during filling.

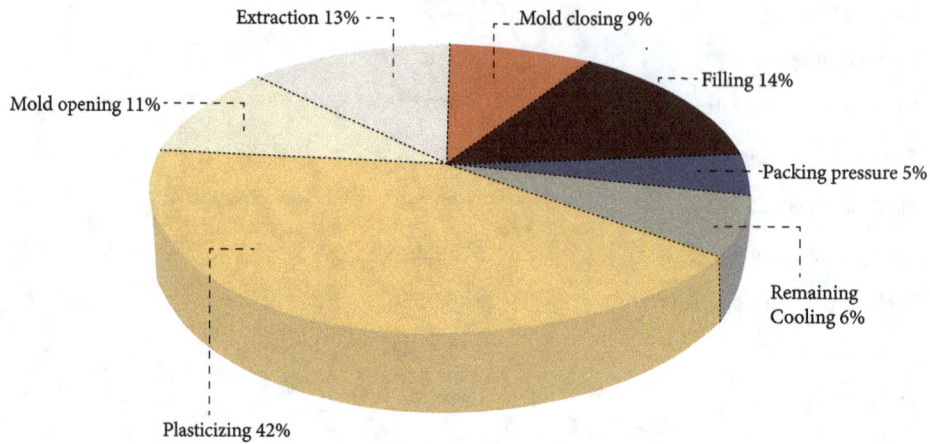

Figure 5.34 Distribution of energy consumption by cycle stages. This is represented in a pie chart. These stages are arranged clockwise in the following order: mold closing, injection, holding pressure, dosing delay, dosing, residual cooling, mold opening, and ejection

5.6.2 Process Intervention

The process was optimized through simulation. The simulation, experimental results, and other calculations allow adjustment of the switching point, injection speed profile, and packing pressure. Table 5.4 presents the results obtained from the simulation. With the new conditions, the cycle time decreased by 13 s with respect to the initial conditions, and the scrap associated with quality defects was reduced from 2% to 1%.

Table 5.4 Final Processing Parameters Resulting from the Intervention

Parameter	Final condition
Cooling time [s]	21
Barrel temperature [°C]	210
Hot runner temperature [°C]	210
Plasticizing time [s]	20
Plasticizing speed [%]	80

Parameter	Final condition
Filling speed profile [%]	45/45/40
Packing pressure profile [kg/m³]	41.5
Packing time [s]	2.5
Cycle time [s]	77.5

5.6.3 Intervention Results

After applying the new processing conditions, the SEC_s was calculated, and the energy gap analysis was applied, with the result shown in Figure 5.35. At the end of the intervention, the SEC_g was reduced from 0.403 kWh/kg to 0.338 kWh/kg, achieving a 16% reduction in energy consumption (see Figure 5.35) with a 15% increase in productivity.

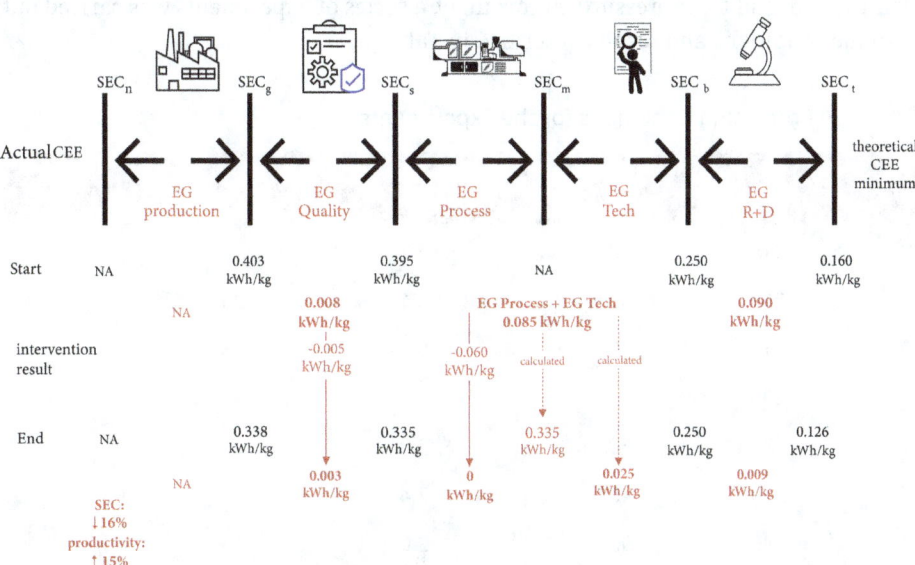

Figure 5.35 Energy gap analysis of the automotive parts injection molding process intervention results. This section presents the values (in black) before and after the intervention for SEC_g, SEC_s, SEC_b, and SEC_t, as well as the respective quality, process + technology, and R&D gaps calculated with these values at the two different time points in the lower central area of the SECs (in red). The results of the intervention are presented (in red) right in the middle of the two time points

The estimated savings, based on a production of 47,000 kg/year, are as follows:

- Energy savings of 3,055 kWh

- For USD 0.10/kWh, the savings represent USD 305/year

- A carbon footprint reduction of 0.47 t CO_2 eq/year, using 0.16438 kg CO_2 eq/kWh conversion factor from hydropower to CO_2 equivalent [1].

Although the energy savings do not seem significant, the productivity improvement would allow an increase from 47,000 kg/year to 54,000 kg/year during the same production time. This represents an increase of 5,385 units. If each unit generates profits of USD 7, this could represent an increase in annual sales of USD 37,695.

5.6.4 Additional Recommendations

The company requested the reduction of energy consumption during plasticization to be explored. To this end, the performance of the process was evaluated at different temperatures and screw rotation speeds, to optimize the melting temperature, plasticizing speed, and back-pressure. Accordingly, a series of experiments was carried out with the conditions and results described in Table 5.5.

Table 5.5 Processing Conditions for the Experiments

Condition	Temperature [°C]	Plasticizing speed [%]	Time difference [s]	Difference in energy consumption [%]
1	215	90	2.2	15
2	215	60	3.1	25
3	215	40	5.1	28
4	230	90	3.9	29
5	230	60	3.9	26
6	230	40	7.4	28
7	245	90	2.0	15
8	245	60	2.7	16
9	245	40	7.9	25

The best energy efficiency was obtained under Condition 6, with a plasticizing speed of 40% and temperatures of 230 °C, as shown in Figure 5.36. However, the plasticizing time is much longer than the cooling time for this speed. This is not convenient, since it implies increasing the cycle time. Therefore, operation under condition 6 is not feasible, since productivity (short cycle times) is usually economically more important than energy consumption. Because of this, the plasticizing speed was kept at 80%.

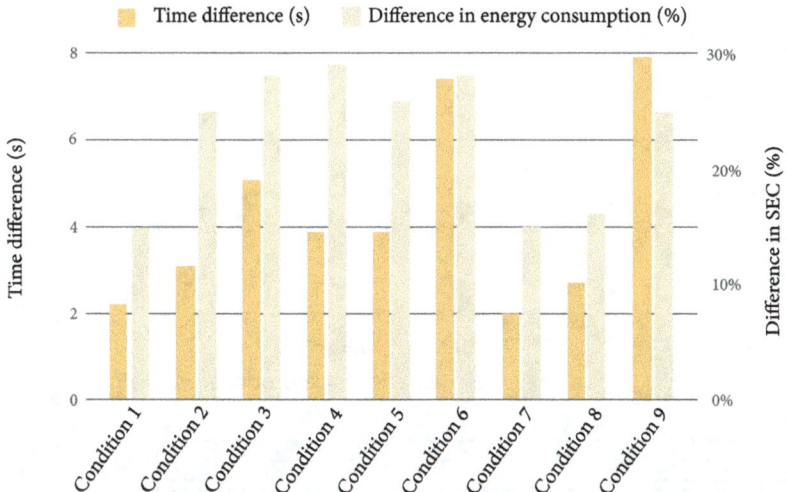

Figure 5.36 Experimental results for proposed parameter conditions for plasticization

Since the energy improvement of plasticization was considered unfeasible, the next gap that could be addressed is a technological one, according to the EGM analysis of the final condition. This implies investment in new equipment, which requires a payback analysis.

5.7 Quality and Production Gap in Extrusion Blow Molding of Personal Care Bottle

5.7.1 Process Diagnosis

A company producing containers for the cosmetic industry was producing a container with the geometry shown in Figure 5.37. It was being produced on continuous extrusion blow molding equipment with a reciprocating station. The mold has two cavities, and the container is produced in a blend of high-density polyethylene (HDPE), low-density polyethylene (LDPE), and a masterbatch pearlescent in linear low-density polyethylene (LLDPE) vehicle. The material parameters are described in Table 5.6. The melt flow index (MFI) was measured at a load of 2.16 kg and a temperature of 190 °C. The cycle time under the usual conditions is 18.5 s, which requires the plasticizing unit to operate at a mass flow rate of 20.8 kg/h and a spindle rotation speed of 48 rpm. Under these conditions, the SEC_s from the data analysis in Figure 5.38 is 0.450 kWh/kg.

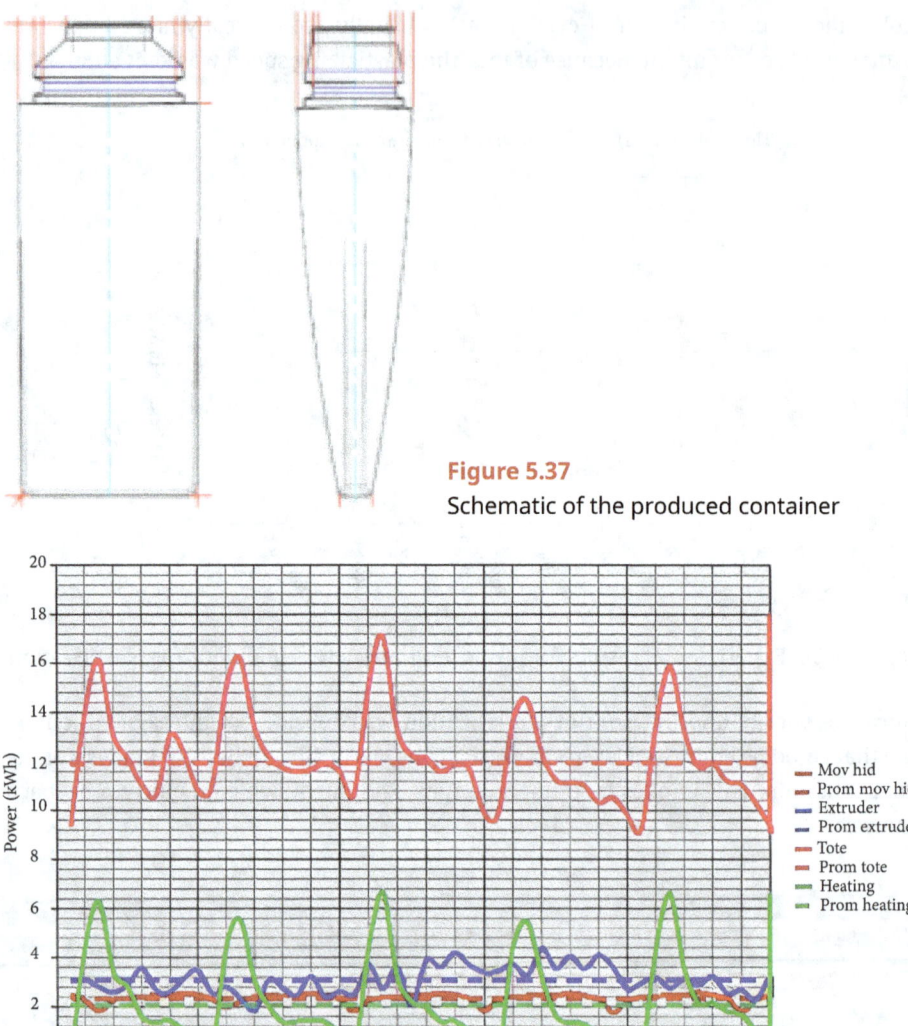

Figure 5.37
Schematic of the produced container

Figure 5.38 Energy measurements of the continuous blow-molding extrusion line under normal process conditions. This is presented in a plot of power [kW] vs. time [s], with the total power shown as a red line and the power of the different parts of the lines in other colors. Lines show the power demanded by the motor for hydraulic movements (Mov hid), the extruder, the totalizer (Tote), and the heating equipment (Heating). Dotted lines depict the average for each variable (named with the word "Prom" before the variable)

Figure 5.39 shows the results of the energy gap analysis for the hollow body blow-molding case study. The combined process and technology energy gap is 0.160 kWh/kg, and the combined production and quality energy gap is 0.650 kWh/kg.

Table 5.6 Material Parameters for Mold Production

Component	MFI (g/10 min)	Density, ρ (g/cm^3)
HDPE	0.3	0.950
LDPE	2	0.918
LLDPE	25	0.924

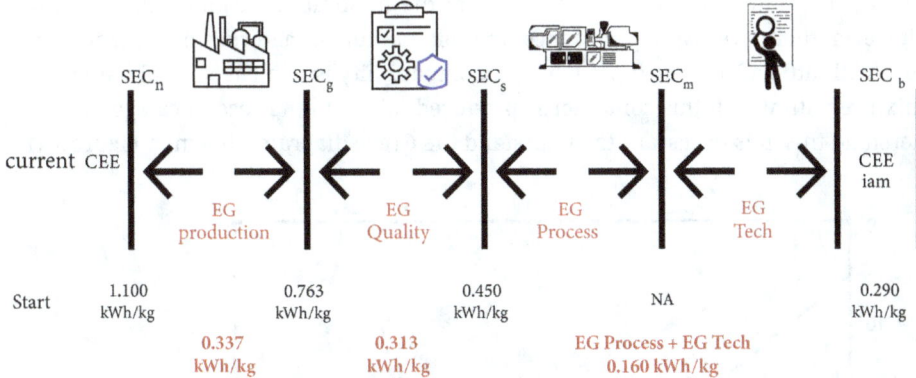

Figure 5.39 Combined process and technology energy gap for the hollow body blow-molding case study. It presents the values (in black) for SEC$_n$, SEC$_g$, SEC$_s$, and SEC$_b$, and the respective quality, process + technology, and R&D gaps calculated with these values in the lower central SEC area (in red)

The energy gap in production and quality is excessively high because about 41% of the parts must be reprocessed due to the appearance of scratches on the surface of the container. This is due to the accumulation of material around the nozzle through which the preform is extruded, known as "die buildup" (Figure 5.40). This dramatically affects the quality of the energy gap. Additionally, this same problem affects the production energy gap since every 20 min, the process must be stopped to clean the nozzle, and the sum of the accumulated downtime is significant.

Figure 5.40
A die buildup problem around the extrusion nozzle. Photo taken from 3M.com

The "die buildup" problem also influences the process energy gap because it forces a reduction in velocity. Material accumulation is intensified when the shear rate at the die wall increases. In this case, the following hypothesis is proposed: if the "die buildup" problem is solved, three energy gaps will be reduced.

5.7.2 Intervention Results

The theory states that one of the causes of die build-up is incompatible mixtures. However, there are two types of incompatible mixtures: materials of a chemical nature with little affinity or those whose incompatibility is of a rheological nature. The mixture with which this container is produced falls into the second category. To understand this, it is necessary to understand the Grace diagram, shown in Figure 5.41.

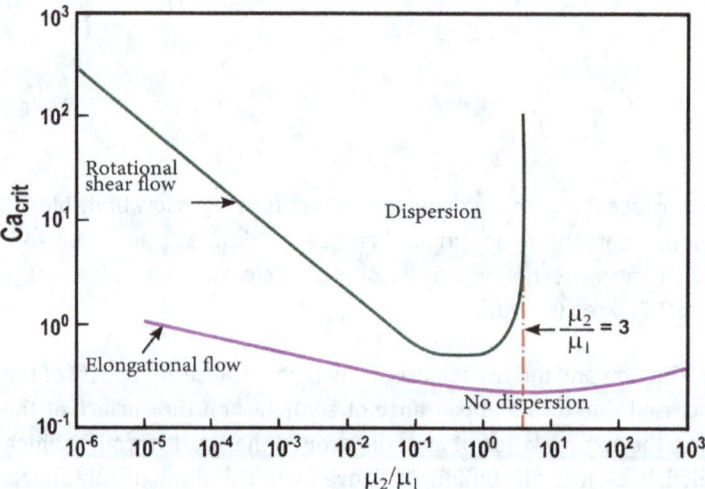

Figure 5.41 Grace diagram, showing the viscosity ratio of the dispersed phase to the continuous phase (μ_2/μ_1) on the x-axis. Elongational flow is shown in purple, and rotational shear flow is shown in green, allowing the dispersion and non-dispersion zones to be observed

In this case, LDPE and LLDPE act as the dispersed phase, while HDPE is the continuous phase, since the material is found in greater proportion in the mixture. On the other hand, the y-axis has the capillary number (C_a). The capillary number is a complex quantity that relates the stress acting on the dispersed phase to the interfacial tension and the diameter of the droplets in the dispersion. However, for the purpose of simplifying the discussion, when C_a increases, the stress that needs to be applied to the mixture to achieve dispersion increases. The purple line represents the C_a limit that must be overcome to achieve dispersion by elongational flows. The green line is the limit that must be overcome to achieve dispersion by shear flows. The limit is much higher and, therefore, more demanding when only shear flows are available. This is the case with a single-screw plasticizing unit of a extrusion blow molding equipment. Elongational flows are negligible, and mixing will depend on exceeding the limit set by the green curve.

A characteristic of the boundary generated by the rotational shear flow is that it presents an area where the stresses required to achieve dispersion are the lowest possible. This occurs when $0.1 \leq \mu_2/\mu_1 \leq 1$. If the mixtures are not in that range, there will be rheological compatibility difficulties and the extrusion equipment will have to provide significantly higher stresses to achieve dispersion. This analysis is performed by obtaining the rheological curves of LDPE, HDPE and masterbatch to evaluate the μ_{LDPE}/μ_{HDPE} and $\mu_{masterbatch}/\mu_{HDPE}$ ratios. However, we will keep the discussion at the qualitative level. A polymer with a lower melt flow index (MFI) is usually more viscous. LDPE and the masterbatch are much more fluid than HDPE. This causes the viscosity ratio to shift to the left in the Grace diagram. These blends fall outside the range recommended by the lower limit.

The mixture of HDPE and LDPE is made to obtain an intermediate stiffness between both materials, according to the requirement for the container's functionality. To solve the problem without affecting the functionality, it was recommended the use of a new mixture between the HDPE (MFI = 0.3 g/10 min, ρ = 0.950 g/cm³) of usual use, with a medium-density polyethylene (MDPE) (MFI = 0.5 g/10 min, ρ = 0.935 g/cm³) and the masterbatch was manufactured using the MDPE as a vehicle. In this way, the viscosities of all materials would be much closer to each other, helping to keep the mixture within the optimal viscosity ratio range to facilitate dispersion.

5.7.3 Intervention Results

Changes in the formulation of the container mix prevented the formation of "die buildup". This reduced the reject rate from 41% to less than 10% and eliminated process stops due to nozzle cleaning.

Regarding the process energy gap, the new formulation reduced the cycle time from 18.5 s to 16 s. This allowed an increase in productivity from 20.8 kg/h to 24.1 kg/h, with an average power demand of 9.4 kWh. The SEC_s under the new condition was 0.390 kWh/kg, which is assumed to be the SEC_m for the machine during the production of the case study package. The result is detailed in Figure 5.42.

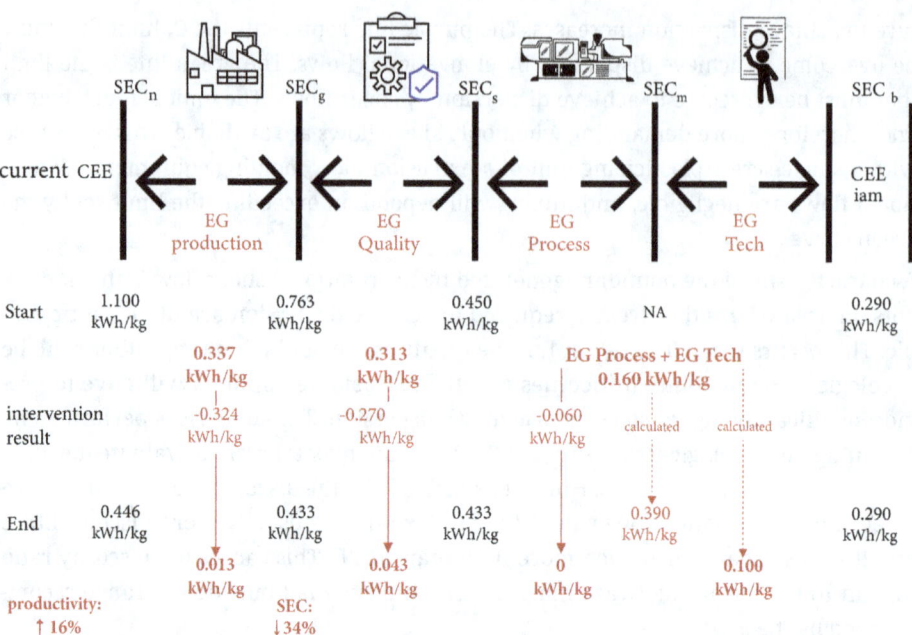

Figure 5.42 Process energy and technology energy gaps after increasing process speed (cycle time reduction). The values (in black) before and after the intervention are presented for SEC_n, SEC_g, SEC_s, and SEC_b, as well as the respective quality, process + technology, and R&D gaps calculated with these values at the two different time points in the lower central area of the SEC_s (in red). The results of the intervention are presented (in red) right in the middle of the two time points

Thus, the main gap closed was the production gap, followed by the quality gap. The intervention reduced specific energy consumption by 34% and increased productivity by 16%.

The estimated savings, based on a production of 150,000 kg/year, are as follows:

- Energy savings of 98,100 kWh

- For USD 0.10/kWh, the savings represent USD 9,810/year

- A carbon footprint reduction of 14.4 t CO_2 eq/year, using 0.16438 kg CO_2 eq/kWh conversion factor from hydropower to CO_2 equivalent [1].

This productivity improvement would allow an increase from 150,000 kg/year to 174,000 kg/year in the same production time. This represents an increase of 24,000 kg. If each unit generates profits of USD 0.25, this could represent an increase in annual profits of USD 6,000.

5.8 Process and Technology Gap in Rubber Injection Molding

5.8.1 Process Diagnosis

A company manufacturing EPDM (ethylene-propylene-diene) rubber parts and profiles had a hydraulic injector, as shown in Figure 5.43. The lower part of the diagram shows the injection unit. The material conveyed by the screw and cylinder system on the right-hand side accumulates in a piston chamber on the left-hand side. When the parts are injected into the mold, the piston moves to generate the necessary pressure for filling.

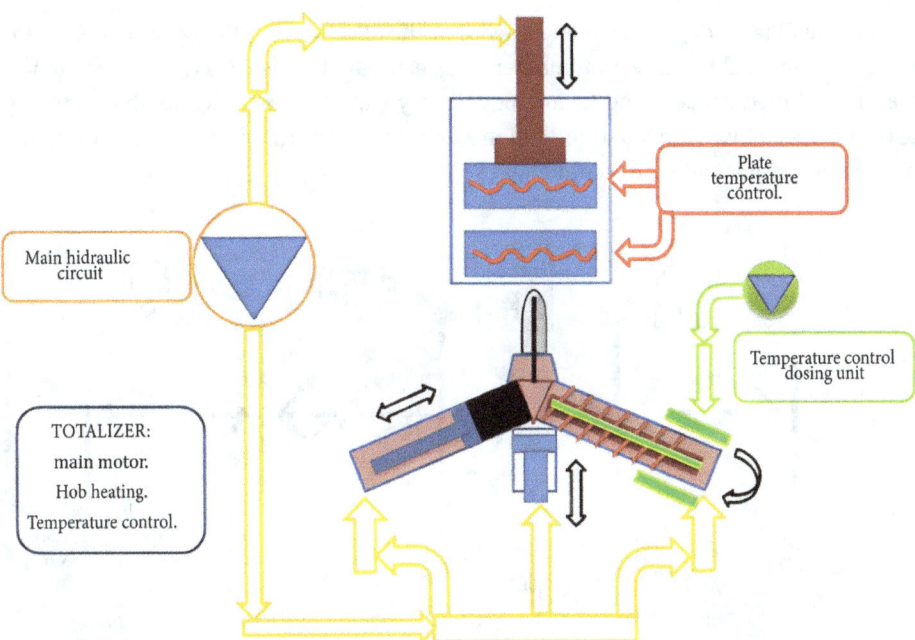

Figure 5.43 Hydraulic rubber injection molding machine

The product with which the diagnosis was performed was a circular part, as shown in Figure 5.44, which was being produced by injecting a 100-cavity mold.

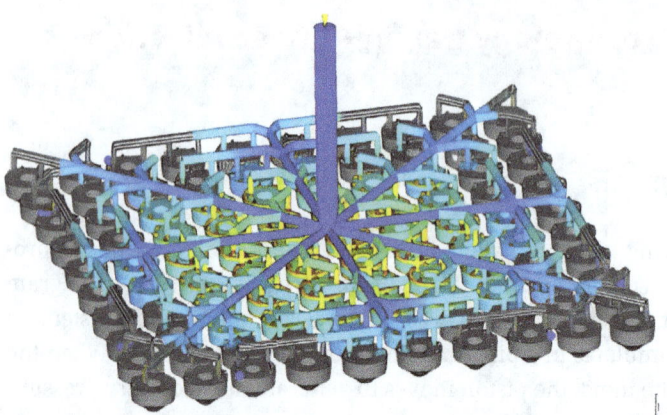

Figure 5.44 Distribution of the parts in the mold

During the diagnostic, it was determined that the SEC_s was 6 kWh/kg, and an SEC_b for state-of-the-art rubber injection molders was estimated to be 0.5 kWh/kg. This generates a combined process and technology energy gap of 5.5 kWh/kg, as shown in Figure 5.45. Special work needed to be done on machine parameterization to close the process gap and isolate the technology gap.

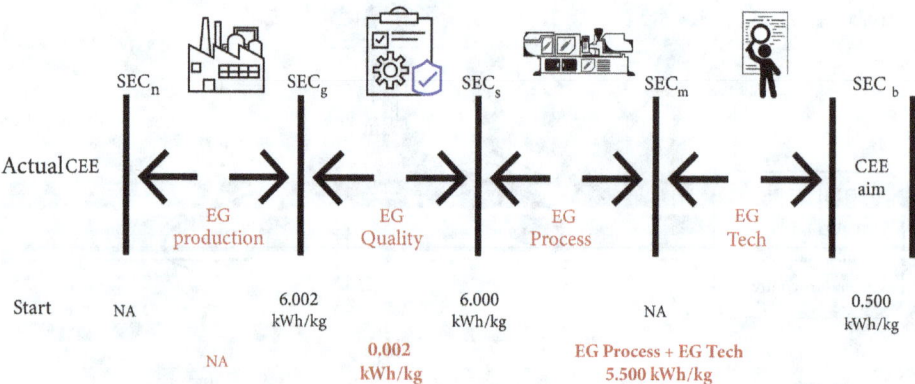

Figure 5.45 Energy gap analysis for the rubber injection line. This section presents the values (in black) for SEC_g, SEC_s, and SEC_b, and the respective quality, process + technology, and R&D gaps calculated with these values in the lower central SEC area (in red)

The main engine's power demand during the injection cycle was monitored to determine the consumption of each stage of the cycle and evaluate the possibility of improvement. The power demand diagram is shown in Figure 5.46. The cycle time is 380 s.

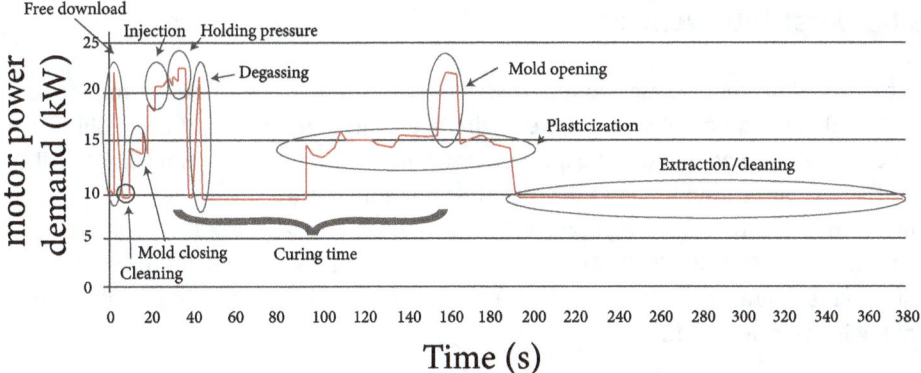

Figure 5.46 Power demand for the rubber part injection cycle. This shows the electrical power profile (red line) in a plot of motor power [kW] vs. time [s] during the injection cycle, with the demarcation of each stage

Figure 5.47 presents the distribution of the power consumption of each of the cycle's stages. It shows that the largest consumptions occur during part extraction, cleaning, and curing. The first is a technological problem associated with mold manufacturing, which will be discussed later. The latter two are a process condition.

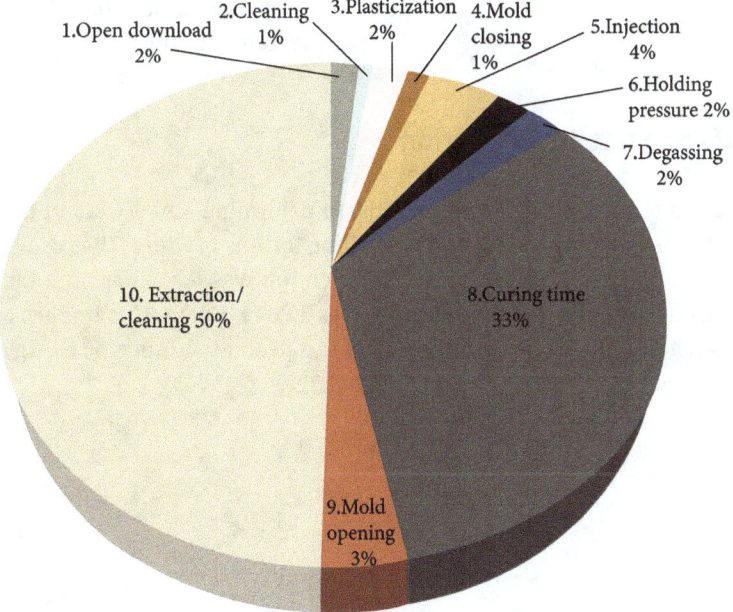

Figure 5.47 The percentage distribution of energy consumption for each stage of the injection cycle, presented as a pie chart

5.8.2 First Intervention

The curing time under these conditions was 180 s. Therefore, the vulcanization kinetics of the rubber compound were studied using a torque rheometer, and the mold-filling process was simulated under the different vulcanization conditions using CADMOULD. For the vulcanization kinetics, the vulcanization temperature was varied and equal to the compound's injection temperature. The temperatures used were 170 °C (actual temperature), 185 °C, and 200 °C. The results are shown in Figure 5.48. By increasing the injection temperature from 170 °C to 200 °C, curing is reduced from 180 s to 60 s, with a saving in cycle time of 120 s.

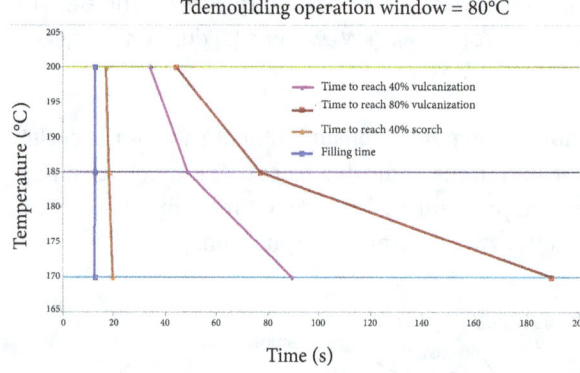

Figure 5.48
Vulcanization kinetics, represented in a plot of time [s] vs. temperature (°C)

5.8.3 Result of the First Intervention

With the new injection temperature, it was possible to determine a new SEC_s in the process, which is assumed to be the SEC_m of the machine for this product. The results are detailed in Figure 5.49. The SEC_s before the intervention was 6 kWh/kg, and the SEC_m determined through the process optimization was 3,760 kWh/kg. This implies a reduction of 2,240 kWh/kg. This intervention resulted in a decrease in specific energy consumption of 37% and an increase in productivity of 32%.

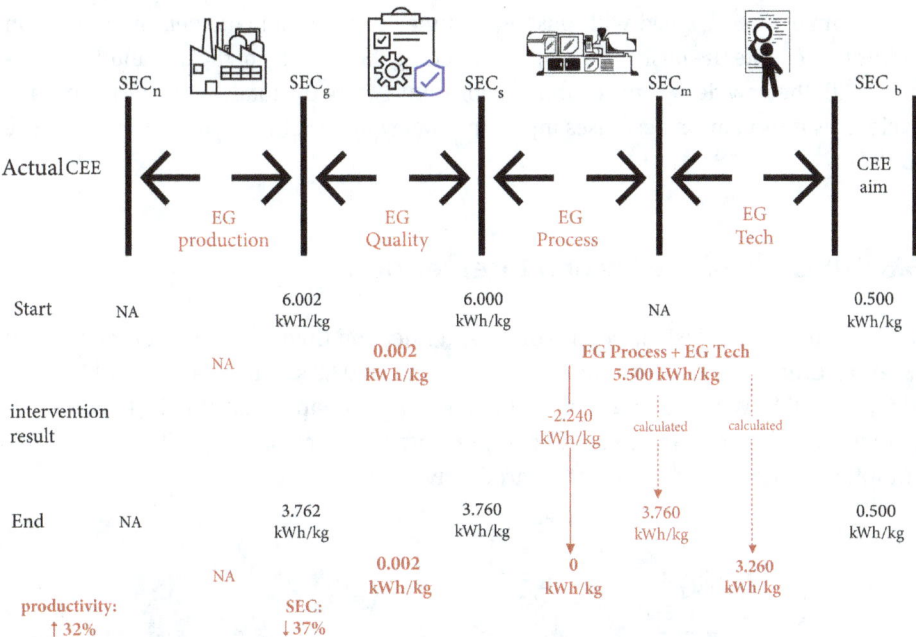

Figure 5.49 Process energy gap and technology energy gap after the injection cycle intervention. This section presents the values (in black) before and after the intervention for SEC$_g$, SEC$_s$, and SEC$_b$, as well as the respective quality, process + technology, and R&D gaps calculated with these values at the two different time points in the lower central area of the SECs (in red). The results of the intervention are presented (in red) right in the middle of the two time points

5.8.4 Second Intervention

The technology energy gap is still very wide. With the curing kinetics intervention, the cycle time was reduced from 380 s to 260 s. Of the improved cycle time, approximately 180 s corresponds to the time it takes to remove and clean parts that remain stuck together in the distribution channels.

The main reason for this extended cycle time is the mold design. Each part has two injection points, but the injection points are too thick, and the mold does not have any runners. Additionally, the distribution channels are also too thick and poorly distributed. Using CADMOULD, the mold filling was simulated using the current channel distribution and geometry of the injection points. The distribution channels account for 25% of the total mass of each cycle, and mold filling takes 7 s. In addition, the geometry of the distribution channels means that the parts that are more toward the center of the mold are filled first, and those at the edges are filled last, as seen in Figure 5.44. This also means that, when post-pressure is applied, the parts in the center tend to be heavier.

A new mold was designed with pushers, thinner distribution channels, and injection points to facilitate the pushers' action of detaching the parts from the distribution channels. With the new design, the channels represent 17% of the total mass injected in each cycle. This intervention decreases injection time by up to 5.3 s, and mold filling is more uniform.

5.8.5 Results of the Second Intervention

With the improved design, savings of 1.7 s in injection time and 165 s in removal and cleaning time were achieved for a total cycle time of 93.3 s. With the new mold, a new SEC_m of 2,550 kWh/kg and a new technology energy gap of 2,050 kWh/kg were obtained, as shown in Figure 5.50. This represented a decrease in specific energy consumption of 32% and an increase in productivity of 64%.

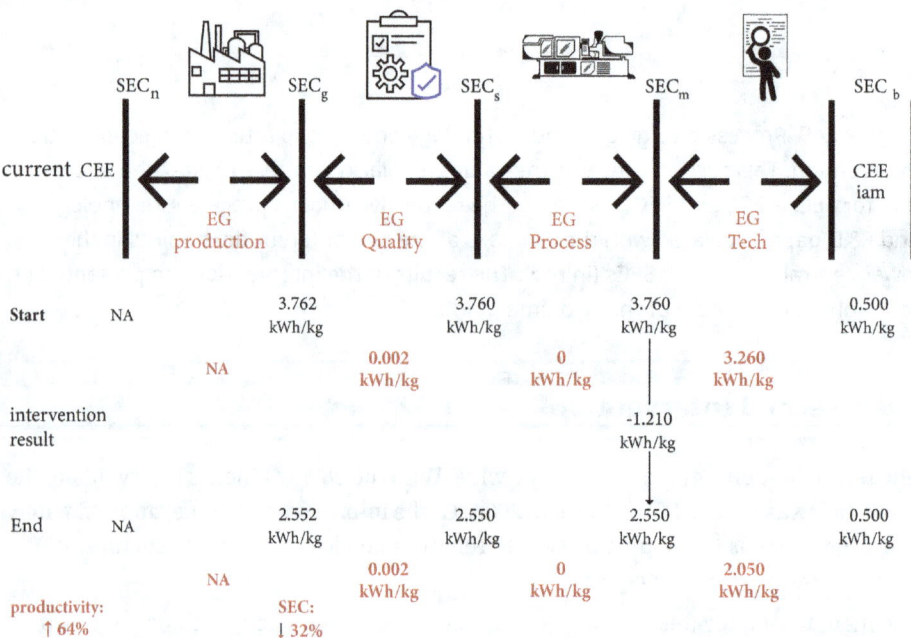

Figure 5.50 New process and technology energy gaps after the injection mold intervention. This graph presents the values (in black) before and after the intervention for SEC_g, SEC_s, SEC_m, and SEC_b, as well as the respective quality, process + technology, and R&D gaps calculated with these values at the two different time points in the lower center of the SECs (in red). The results of the intervention are presented (in red) right in the middle of the two time points

The final result of the two interventions represented a decrease in specific energy consumption of 58% and an increase in productivity of 216%. This translated into an increased production capacity of four times the initial production.

The estimated savings, based on a production of 170,000 kg/year, are as follows:

- Energy savings of 586,500 kWh

- For USD 0.10/kWh, the savings represent USD 58,650/year

- A carbon footprint reduction of 85.9 t CO_2 eq/year, using 0.16438 kg CO_2 eq/kWh conversion factor from hydropower to CO_2 equivalent [1].

This productivity improvement would allow an increase from 170,000 kg/year to 368,016 kg/year in the same production time. This represents an increase of 198,016 kg. If each kg generates sales of USD 1, this represents an increase in annual profits of USD 198,016.

5.9 Process and Quality Gap in Thermoforming Sheet Extrusion

5.9.1 Process Diagnosis

An electronics and appliance manufacturer had a thermoforming sheet co-extrusion line, as shown in Figure 5.51. The outer layer extruders place 2.5% thick layers of crystal polystyrene (PS), and the center extruder places a layer of high-impact polystyrene (HIPS) at 90% of the thickness distribution. About 33% of the thermoformed product is discarded due to quality problems. In the usual process condition, the co-extrusion system operates with a mass flow rate of 374 kg/h, producing sheets 811 mm wide and 1.5 mm thick.

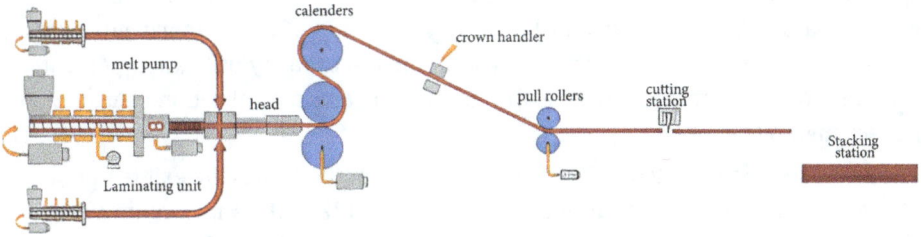

Figure 5.51 Co-extrusion line for the thermoforming film

When energy consumption was studied using the EGM, the SEC_s was found to be 0.556 kWh/kg. According to reference data obtained at international fairs, a line of this type should have an SEC$_b$ close to 0.350 kWh/kg. With these values, the combined process and technology gap is 0.206 kWh/kg, as shown in Figure 5.52. The highest energy gap is the quality gap, with a value of 0.274 kWh/kg, which is explained by the level of reprocessing due to quality problems in thermoforming. However, in this case, the quality energy gap was the most difficult point at which to intervene, since the causes could be associated with the process or technology. Most of the time, when interventions are made with the process and technology energy gap, significant improvements in quality are also obtained. Therefore, it was decided to work primarily on the parameterization of the machines to close the process gap and isolate the technology gap.

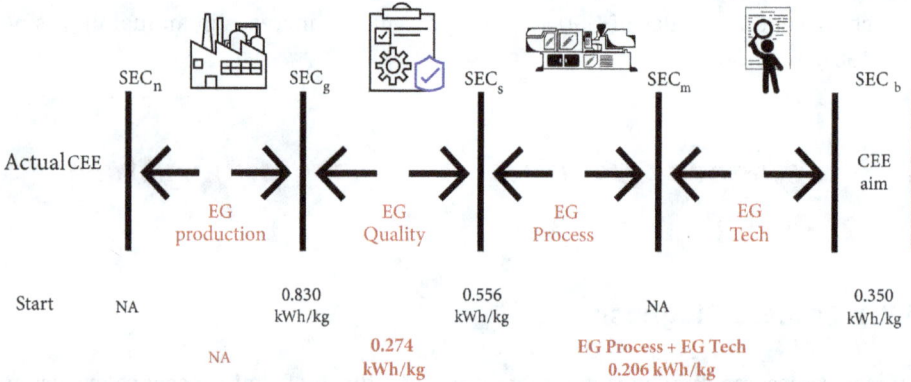

Figure 5.52 Combined process + technology energy gap of the sheet extrusion line. This shows the values (in black) for SEC$_g$, SEC$_s$, and SEC$_b$, and the respective quality, process + technology, and R&D gaps calculated with these values in the lower central SEC area (in red)

In analyzing the process to reduce the process gap, it was identified that the production line operates with a melt pump that regulates the rotational speed of the main extruder screw. In addition, the extruder has a 1760 mm broad flat die. Table 5.7 presents the parameters of the standard operating condition of the main extruder, which represents 95% of the line structure. In addition, the capacity utilization percentage column is obtained from evaluating the equipment capacity utilization using the bottleneck method.

The gap analysis presented indicates that the main bottleneck is the exit temperature of the sheet in the calender frame, which reaches 122% of its maximum value. To avoid problems, the sheet must exit the calender system at least 10 °C below the glass transition temperature (T_g). In the case of HIPS, the T_g is 100 °C. An exit temperature above this threshold can induce a reorientation in the sheet due to the dragging of the pulling rolls, which causes shrinkage during thermoforming and hinders the process.

Table 5.7 Parameter Values of the Usual Operating Condition of the Sheet Co-extrusion Line and Bottleneck Analysis

Variable	Current value	Capacity	Capacity utilization [%]
Power demand of main motor [kW]	60	225	26.7
Screw rotational speed [rpm]	50.7	155	32.7
Heating power [kW]	32	66.5	48.1
Maximum speed of the melt pump [rpm]	70	82	85.4
Calender speed [m/min]	4.5	6.2	72.6
Puller speed [m/min]	4.5	10.2	44.1
Melt pressure [bar]	1,730	5,000	34.6
Sheet width [mm]	811	1,720	47.2
Calender outlet sheet temperature [°C]	110	90	122

In second place in percentage of utilization is the melt pump speed (85.1%), and in third place is the calender speed (72.6%). All other variables are used below 50%. The most concerning value is the die width, which is used at 47%. The line could produce a sheet twice as wide, and it could generate a higher flow to fill the entire head if cut it in half. The bottleneck analysis indicates that special emphasis should be placed on the calender frame and the melt pump.

The company recently acquired the melt pump to improve sheet quality. Paradoxically, it was not purchased based on the capacity of the main extruder but according to historical production conditions. Thus, the melt pump has become a constraint to better quality and higher productivity.

5.9.2 Process Intervention

A test by decoupling the melt pump showed that the main plasticizing unit could handle a mass flow rate of more than 900 kg/h. The test was done using the entire die width, but it was confirmed that there was insufficient capacity in the calender system to cool this flow. A detailed analysis of the calender system showed that the diameter and speed of the calenders were insufficient. The calender rolls had a diameter of 311 mm, and for the plasticizing unit's capacity, they should be at least 800 mm and have a speed of 10.2 m/min.

The following were the proposed changes:

- Eliminate the melt pump (zero cost)

- Purchase a new calender system (with an approximate cost of USD 200,000).

5.9.3 Intervention Results

With this intervention, double sheets at a processing speed of 850 kg/h and a reduction in thermoforming reprocessing below 10% were possible under the operating conditions presented in Table 5.8. These conditions better utilized the capacities of the sheet extrusion line.

Table 5.8 Operating Condition Values after the Intervention in the Sheet Co-extrusion Line and Corresponding Bottleneck Analysis

Variable	Current value	Capacity	Capacity utilization [%]
Power demand of main motor [kW]	115	225	51.1
Screw rotational speed [rpm]	136	155	87.7
Heating power [kW]	55	66.5	82.7
Maximum speed of the melt pump [rpm]	Not available	Not available	Not available
Calender speed [m/min]	10.2	10.2	100
Puller speed [m/min]	10.2	10.2	100
Melt pressure [bar]	2,400	5,000	48.0
Sheet width [mm]	1,622	1,720	94.3
Calender outlet sheet temperature [°C]	85	90	94.4

Figure 5.53 presents an analysis of the energy gaps of the sheet extrusion line after the intervention. As previously mentioned, the intervention in a lower-level gap has additional effects on the higher-level gaps. The process gap is closed with the results achieved, and the technology gap remains at a low value of 0.056 kWh/kg. The reduction of the process gap is 0.150 kWh/kg. Additionally, the quality energy gap, which was 0.274 kWh/

kg, was reduced to 0.045 kWh/kg, implying a reduction of this gap by 0.229 kWh/kg. This represents a decrease in specific energy consumption of 46% and an increase in productivity of 205%.

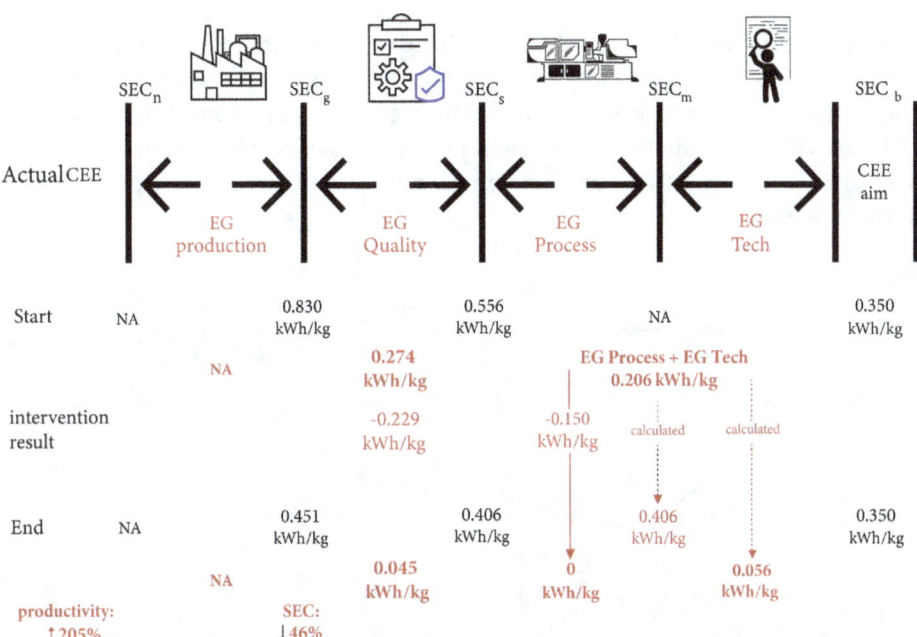

Figure 5.53 Process energy gap and technology energy gap after the sheet extrusion line upgrade. This graph shows the values (in black) before and after the intervention for SEC_g, SEC_s, and SEC_b, as well as the respective quality, process + technology, and R&D gaps calculated with these values at the two different time points in the lower central area of the SECs (in red). The results of the intervention are presented (in red) right in the middle of the two time points

The estimated savings, based on a production of 3,500,000 kg/year, are as follows:

- Energy savings of 1,326,500 kWh

- At USD 0.10/kWh, the savings represent USD 132,650/year

- A carbon footprint reduction of 2.6 t CO_2 eq/year, using 0.16438 kg CO_2 eq/kWh as the conversion factor from hydropower to CO_2 equivalent [1].

This productivity improvement would allow an increase from 3,500,000 per year to 7,175,000 kg/year in the same production time. This represents an increase of 28,700 kg if each kg generates profits of USD 0.25, or an annual profit of USD 918,750.

5.10 Process Gap in Blown Film Extrusion

5.10.1 Process Diagnosis

In a blown film extrusion plant, there is a single-layer extruder with internal bubble cooling (IBC), 3.5 in screw, and $L/D = 25$. The screw is a barrier screw with Saxton-type distributive mixer at the tip, and which can rotate at a maximum speed of 120 rpm. The geared motor is 40 kW. A schematic image is presented in Figure 5.54. A LLDPE-C4 film of 1.64 m tubular width and 0.37 mils thickness was produced during the tests.

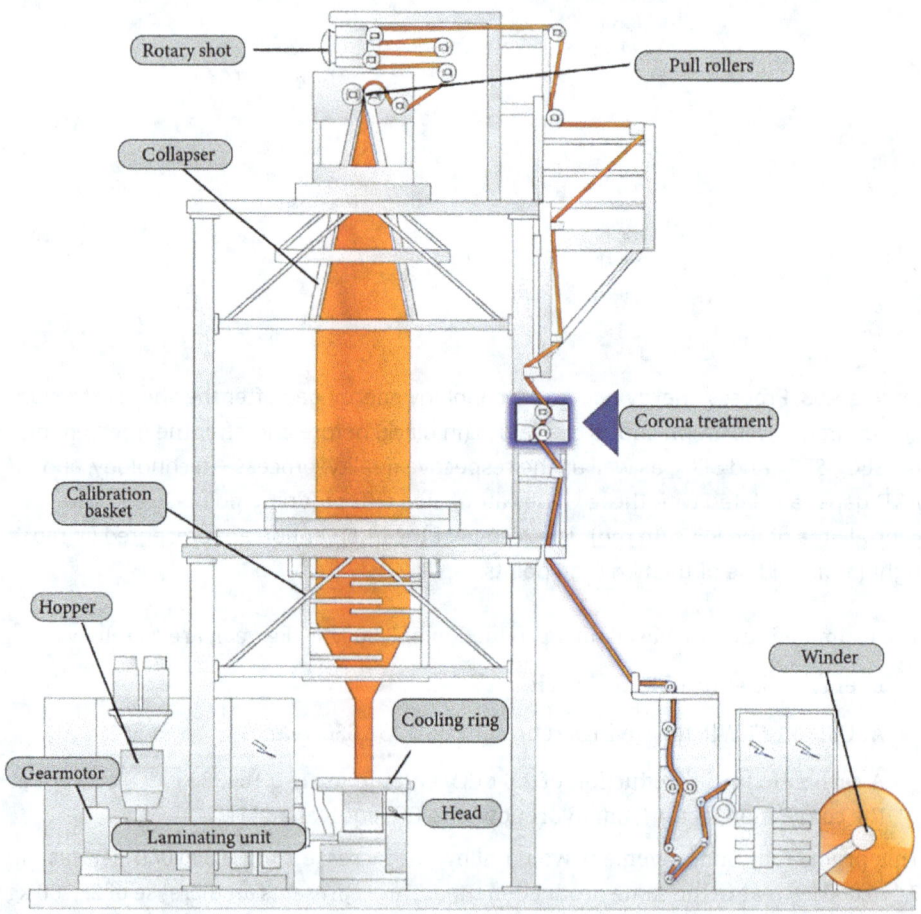

Figure 5.54 Typical monolayer blown film extrusion line

A power meter was connected to the totalizer and another to the machine's main motor. The results of the measurements performed are shown in Figure 5.55. Studying energy consumption using the EGM yielded the result shown in Figure 5.56. Although it is a small gap, closing the process gap is always advisable because it means higher productivity, and the economic impact of the improvement is usually significant for the company's competitiveness. The work therefore needed to focus on the parameterization of the machines to close the process gap and isolate the technology gap.

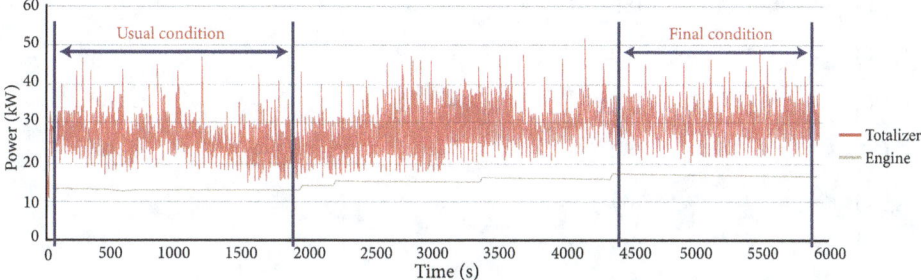

Figure 5.55 After closing the process gap, power demand measurements for the totalizer and prime mover under normal process conditions and at the final condition. The profile of totalized electrical power (red line) and the motor power (brown line) is presented in a plot of time [s] vs. power [kW]

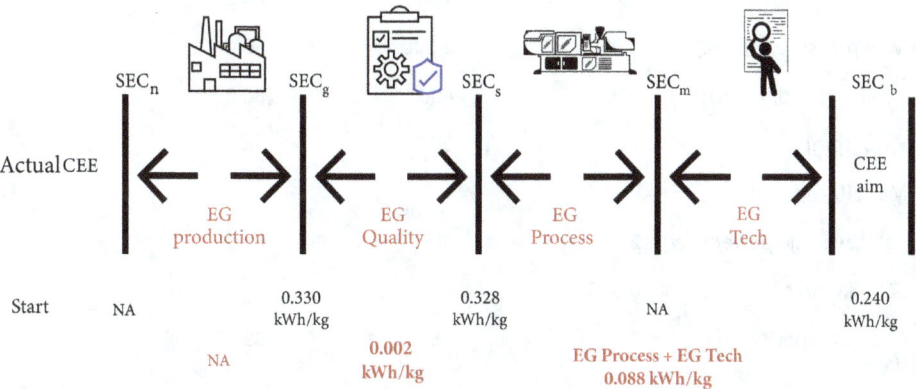

Figure 5.56 Energy gap analysis of the blown film extrusion line. This section presents the values (in black) for SEC_g, SEC_s, and SEC_b, and the respective quality, process + technology, and R&D gaps calculated with these values in the lower central SEC area (in red)

Table 5.9 presents the parameters of the typical process conditions and the results of a bottleneck analysis. The main bottlenecks are melt temperature (96.3%) and cooling capacity (89.7%). Otherwise, there are no apparent restrictions on pulling speed, spindle rotation speed, primary motor power demand, or melt pressure.

Table 5.9 Typical Processing Conditions, Maximum Capacity of the Main Equipment Components, and the Primary Process Constraints and Bottleneck Assessment

Variable	Typical value	Max capacity	Use of max. capacity [%]
T_1 [°C]	156	—	—
T_2 [°C]	160	—	—
T_3 [°C]	175	—	—
T_4 [°C]	180	—	—
T_5 [°C]	185	—	—
$T_{\text{screen block}}$ [°C]	185	—	—
$T_{\text{neck adapter}}$ [°C]	185	—	—
T_{block} [°C]	185	—	—
Melt temperature, T_{melt} [°C]	183	190	96.3
N [rpm]	29	120	24.2
Melt pressure, P_{m} [psi]	4,000	5,000	80.0
Pulling speed, V_{h} [ft/min]	320	400	80.0
V_{doser} [rpm]	14	30	46.7
Mass flow [kg/h]	82.4	—	—
Total average power [kW]	27	40	67.5
SEC_{s} [kWh/kg]	0.328	—	—
Cooling capacity, Q_{cool} [kW]	14.3	16	89.7
Average gear motor power [kW]	14	30	46.7

The blown film extrusion process inherently requires significant cooling of the film. If the melt is not cooled properly and quickly, the bubble is destabilized, and it is not easy to control the thickness distribution of the final product. If the processing speed is to be increased, either the cooling capacity must be increased (in which case the solution becomes technological and thus becomes a technology gap), or the cooling requirements must be reduced. Increasing the process speed increases the cooling requirements.

The possibility of exiting the plasticizing unit at a significantly lower melt temperature, without affecting extrudate quality, must be evaluated to reduce cooling needs. For this purpose, an analysis of the screw geometry and its correspondence with the heating zones of the plasticizing unit was performed. This is presented in Figure 5.57.

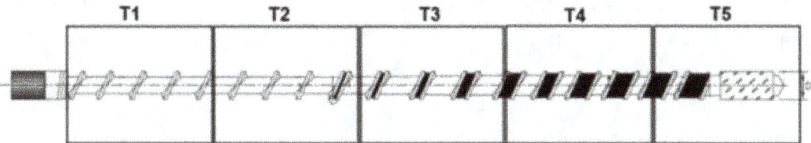

Figure 5.57 Representation of the screw and its correspondence with the heating zones of the blown film extrusion line

The screw has a long feeding zone. The barrier, which is designed so that plasticization starts near its beginning, is located towards the end of the heating zone T_2. On the other hand, LLDPE has a melting temperature of 122 °C, and a target mass temperature can be found at around 170 °C. It is also important to consider that the melt tends to heat up more due to viscous dissipation and residence time.

5.10.2 Process Intervention

A decision was made to lower T_1 and significantly increase T_2 to ensure that the material reaches the start of the barrier above the melting temperature. T_3, T_4, and T_5 were also lowered, leaving the highest temperature in the mixer region where a significant flow restriction is generated. Likewise, the temperatures of the zones corresponding to the adapter, the feed block, and the screen changer were reduced, considering that the plasticization and homogenization work must have already taken place in the screw. The new operating conditions are presented in Table 5.10.

Table 5.10 Processing Conditions after the Intervention, with Maximum Capacity for the Main Components of the Equipment and Main Process Restrictions for Evaluating Bottlenecks

Variable	Typical value	Max capacity	Use of max. capacity [%]
T_1 [°C]	140	—	—
T_2 [°C]	180	—	—
T_3 [°C]	170	—	—
T_4 [°C]	175	—	—
T_5 [°C]	180	—	—
$T_{screen\ block}$ [°C]	175	—	—
$T_{neck\ adapter}$ [°C]	175	—	—
T_{block} [°C]	175	—	—
Melt temperature, T_{melt} [°C]	167	190	87.9
N [rpm]	30	120	25.0
Melt pressure, P_m [psi]	4,470	5,000	89.4
Pulling speed, V_h [ft/min]	400	400	100.0
V_{doser} [rpm]	17	30	56.7
Mass flow [kg/h]	107	—	—
Total average power [kW]	30.8	40	77.0
SEC_s [kWh/kg]	0.288	—	—
Cooling capacity, Q_{cool} [kW]	14.86	16	92.9
Average gear motor power [kW]	18	30	60.0

5.10.3 Intervention Results

With the process changes made, it was possible to determine a new SEC_s in the process, which is assumed to be the SEC_m of the machine for this product. Figure 5.58 presents the analysis of the energy gaps after the intervention in the extrusion process. The process gap is closed by intervening and adjusting the production speed under the new conditions. The SEC_s before the intervention was 0.328 kWh/kg and the SEC_m determined through process optimization was 0.288 kWh/kg. This implies a reduction of 0.040 kWh/kg. Going from 82.4 kg/h to 107 kg/h implies a 24.6 kg/h increase in productivity. Thus, the intervention results in a decrease in specific energy consumption of 12% and an increase in productivity of 30%.

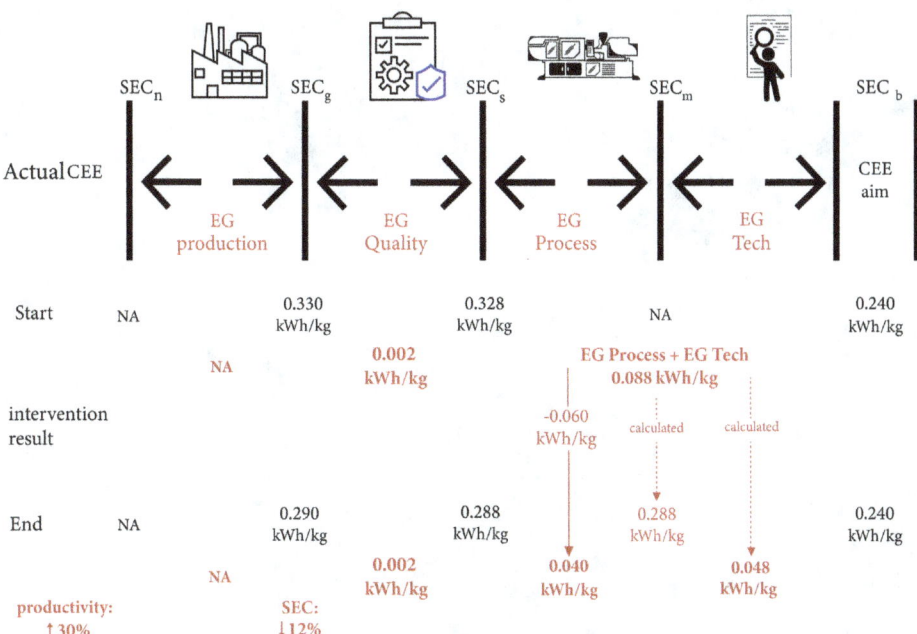

Figure 5.58 Analysis of the energy gaps after the intervention in the extrusion process. This section presents the values (in black) before and after the intervention for SEC_g, SEC_s, and SEC_b, as well as the respective quality, process + technology, and R&D gaps calculated with these values at the two different time points in the lower central area of the SEC_s (in red). The results of the intervention are presented (in red) right in the middle of the two time points

The estimated savings, based on a production of 800,000 kg/year, are as follows:

- Energy savings of 32,000 kWh

- At USD 0.10/kWh, the savings represent USD 3,200/year

- A carbon footprint reduction of 4.69 t CO_2 eq/year, using 0.16438 kg CO_2 eq/kWh conversion factor from hydropower to CO_2 equivalent [1].

This productivity improvement would allow an increase from 800,000 kg/year to 1,040,000 kg/year in the same production time. This represents an increase of 204,000 kg if each kg generates profits of USD 0.25. This could represent an increase in annual profits of USD 60,000/year.

References

[1] "Factor de emisión de CO$_2$ en la generación eléctrica de Colombia", XM, *https://www.xm.com.co/noticias/en-colombia-factor-de-emision-de-co2-por-generacion-electrica-del-sistema-interconectado* [accessed 29 March 2025]

6

Industry 4.0 Enabling Technologies

Julián Patiño, Carlos Correa, Omar Estrada

This chapter explores how Industry 4.0 technologies, such as the "internet of things" (IoT), artificial intelligence (AI), and cloud computing, can improve energy efficiency and production in the polymer processing industry. It presents a multi-layered architecture that includes real-time data acquisition using smart sensors, secure communication over industrial networks, big data storage and analysis, and AI models' application for process optimization and predictive maintenance. Furthermore, integrating these systems with enterprise management software (ERP, EMS, SCADA) for efficient production planning and execution is highlighted. A case study in a plastics manufacturing company not only demonstrates but also inspires, with the successful implementation of these technologies significantly reducing energy consumption and improving productivity.

6.1 Introduction

The advent of Industry 4.0 marks a transformative era in manufacturing, characterized by integrating advanced digital technologies into industrial processes. This fourth industrial revolution encompasses innovations such as IoT, big data analytics, AI, and cyber-physical systems, all converging to create smart factories with enhanced connectivity and automation. These advances offer significant potential to optimize energy consumption, reduce waste, and improve overall operational performance [1], while enabling improved energy efficiency and production effectiveness across various economic sectors, including the polymer processing industry. Figure 6.1 represents the functional architecture of a Cyber-Physical System (CPS), a core element of Industry 4.0, showing the dynamic integration between the physical world, the cyber world, and their communication through a network layer [2]. This diagram illustrates how

Industry 4.0 enabling technologies could facilitate the optimization of energy and production efficiency through real-time feedback and autonomous control.

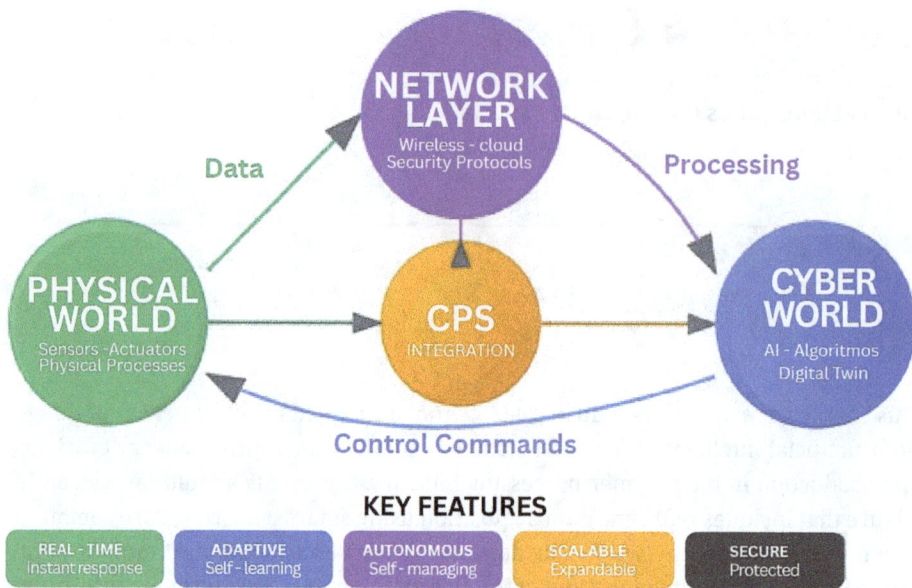

Figure 6.1 Architecture of a cyber-physical system (CPS) in the context of Industry 4.0,
demonstrating the interaction between the physical world, the network layer, and the
cyber world. This architecture enables real-time control capabilities, analytics based
on artificial intelligence and digital twins, and autonomous decision-making aimed at
energy and production efficiency

Polymer processing industries are highly energy-intensive, with significant electricity
and thermal energy requirements for process such as extrusion, injection molding, and
blow-molding processes, and others [3]. Conventional manufacturing methods often involve substantial energy use, rising operational costs, and increasing environmental
impact. Traditional approaches to energy management rely on historical data and periodic audits, which usually fail to provide real-time insights into inefficiencies. The advent of Industry 4.0 technologies enables continuous real-time monitoring and adaptive control of energy usage, leading to cost savings and sustainability improvements,
reducing waste and enhancing overall efficiency [4]. The research indicates that the
adoption of advanced manufacturing robotics (AMRB), advanced visualization and artificial intelligence (AVAI), big data and analytics (BDAA), and IoT technologies posi-
tively impact variables such as productivity, manufacturing errors, product innovation,
decision-making, and process control [5]. Also, implementing sensors, data acquisition
systems, and interconnected networks allows for precise energy tracking and anomaly
detection [6]. This technological shift gives manufacturers real-time feedback on en-

ergy consumption, facilitating predictive maintenance, process optimization, and waste minimization [7].

In the context of polymer processing, the application of these technologies can lead to significant advances. For example, in extrusion-related polymer processing, energy efficiency can be improved by adopting advanced monitoring and control systems that optimize process parameters, reducing energy consumption and enhancing product quality [3]. Key Industry 4.0 technologies that play a role in improving energy and production efficiency in polymer processing include:

- IoT and industrial IoT (IIoT): Protocols that enable interconnectivity between machines, sensors, and control systems for real-time monitoring tasks

- Data acquisition systems: Systems that collect and transmit energy and operational data for analysis and optimization

- Big data management and storage: Practices that ensure efficient handling of vast operational and energy-related data

- Data analytics and AI: Tools used to identify patterns and inefficiencies, enabling automated decision-making

- Information systems integration: Systems that connect enterprise resource planning (ERP) and manufacturing execution systems (MES) with production lines to achieve synchronized energy management.

Despite its benefits, the adoption of Industry 4.0 in polymer processing faces several challenges, including high initial investment costs, cybersecurity concerns, and the need for workforce upskilling [8]. Companies must develop strategic roadmaps that include:

- Technology assessment: Evaluating the readiness of existing infrastructure for digital transformation

- Integration strategies: Ensuring compatibility between legacy systems and recent technologies

- Data governance: Establishing secure data collection, storage, and analysis protocols

- Training programs: Equipping personnel with the necessary skills to leverage Industry 4.0 tools effectively.

Implementing Industry 4.0 in polymer processing requires a structured approach that aligns technological advances with business objectives and sustainability goals. This chapter will examine specific Industry 4.0 technologies, exploring a layered approach to describing Industry 4.0 implementations with a focus on energy efficiency for polymer processing. Sections will show their applications, commonly used tools, challenges, and best practices. By understanding these technologies and their implementation strategies, polymer industry manufacturers can make informed decisions to enhance productivity, minimize energy consumption, and improve operational performance and sustainability.

6.2 Multi-Layer Architecture for Industry 4.0

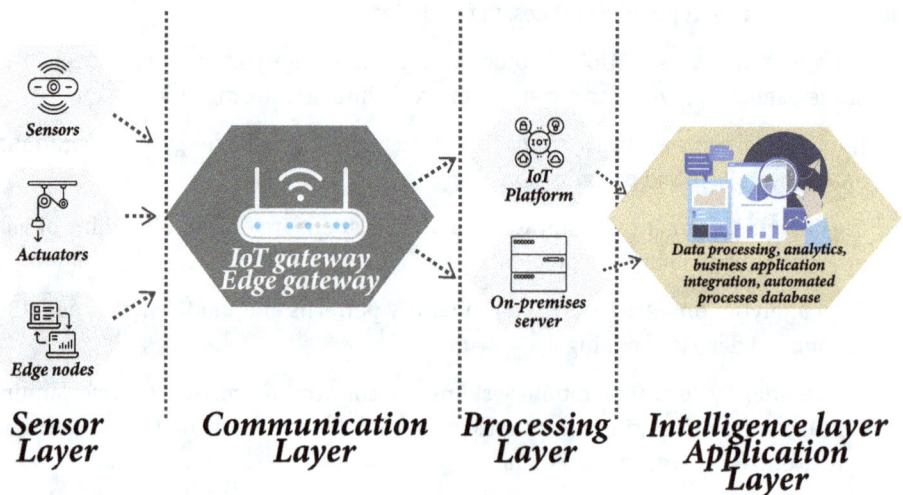

Figure 6.2 Multi-layer architecture model for Industry 4.0-driven energy efficiency in polymer processing. The framework integrates physical, cyber, and decision-making layers to enable real-time monitoring, predictive analytics, and optimization of energy usage in production systems

The successful implementation of Industry 4.0 in energy efficiency for polymer processing relies on a multi-layer architecture that systematically integrates data acquisition, industrial communication, cloud and edge computing, data analytics, and enterprise systems. This structured approach ensures that real-time monitoring, predictive analytics, and automated decision-making are seamlessly executed across all production levels [9]. Manufacturers can achieve digital transformation while minimizing operational costs. This section details the five primary layers of the Industry 4.0 architecture for polymer processing, each with distinct roles and enabling technologies. Figure 6.2 describes the five main layers that our experience can identify:

- Sensor layer: Captures real-time data using sensors and actuators

- Communication layer: Transmits information using industrial protocols such as MQTT and OPC-UA

- Processing layer: Stores and analyzes information on local servers or cloud platforms

- Intelligence layer: Applies AI algorithms for energy optimization and predictive maintenance

- Application layer: Integrates processed data into enterprise systems such as ERP, MES, and SCADA (supervisory control and data acquisition), and enables efficient polymer production energy management.

6.2.1 Sensors and Devices Layer

The sensor layer is crucial in acquiring real-time data from the production floor, serving as the foundation for energy-efficient strategies in polymer processing. This layer collects critical operational data from energy-intensive equipment such as extruders, injection molding machines, and blow-molding units. By continuously monitoring variables such as energy consumption, material usage, and production efficiency, manufacturers can implement proactive measures to reduce waste and optimize performance [10].

A key component of this layer is the intelligent sensor network, which measures essential parameters such as temperature, pressure, power consumption, and material properties. These sensors provide precise data for energy monitoring and process control, ensuring that machines operate within optimal conditions. In addition to traditional sensors, sensorless measurement systems have gained prominence. These AI-driven technologies estimate critical process variables, such as torque, viscosity, and melt flow index, without requiring physical sensors, reducing hardware costs and maintenance requirements [10]. Energy meters and actuators are also integral to this layer, offering real-time power consumption data and enabling adaptive energy management by adjusting machine settings dynamically.

The components in this layer must meet strict technical requirements, including high-resolution data acquisition, low latency, and robust industrial communication protocols, to ensure high performance and accuracy. This section explores the technical specifications, challenges, and best practices for implementing an efficient sensor layer in polymer manufacturing. Standardized communication protocols ensure the seamless transmission of sensor data across industrial networks [11].

Implementing a robust sensor layer offers several key benefits for energy efficiency in polymer processing. This layer helps minimize unplanned downtime by enabling predictive maintenance, ensuring that machinery operates efficiently [10]. Real-time power consumption monitoring reduces energy waste by identifying inefficiencies and enabling corrective actions [3]. Moreover, optimizing process parameters through advanced data acquisition leads to higher production efficiency, lower operational costs, and improved sustainability. As polymer manufacturing transitions toward fully digitalized operations, the sensor layer will continue to serve as the foundation for intelligent energy management. Future advances in AI-driven sensor fusion, 6G-enabled IoT networks, and blockchain-secured industrial data will further enhance the capabilities of the sensor layer, which could impact energy efficiency in polymer processing..

6.2.1.1 Key Components

Smart sensors for real-time data acquisition

Smart sensors are the backbone of Industry 4.0. In real time, they capture physical parameters such as temperature, pressure, torque, power consumption, vibration, and material properties [11]. These sensors provide high-resolution data for energy optimization and quality control. Depending on their intended function, sensors have various characteristics and are used in multiple applications.

In the plastics industry, implementing energy-efficient practices relies on sensors and devices that enable real-time monitoring of operating conditions [10, 11]. Sensors such as thermocouples are essential for measuring temperatures in extrusion or injection molding processes, ensuring thermal stability throughout the operation. Energy consumption sensors capture data on electricity usage in machinery, optimizing resource use. Additionally, production data sensors, including those measuring operating speed, the number of manufactured pieces, and cycle time, play a crucial role in improving operational efficiency.

Microcontrollers from various manufacturers and architectures, such as ATMEL, ARM, ESP32, and PLCs (programmable logic controllers), are intermediaries that capture and process this data, integrating with different communication systems. These devices can digitize, store, and process sensor data, making it available for local use or transmission to other devices. Figure 6.3 illustrates the fundamental components of the sensor layer and control elements (for processing, executing, and making decisions on the acquired data) in industrial environments, especially applicable to polymer processing. The most common types of sensors in polymer processing are:

- Temperature sensors (thermocouples, RTDs, infrared sensors): Used to monitor extruder barrel temperatures, mold heating elements, and polymer melt temperatures

- Pressure sensors: Used to measure internal pressure in extruders, injection molds, and compressed air systems

- Power meters (smart energy monitors): Used to track electricity consumption at the machine level to detect inefficiencies

- Vibration and acoustic sensors: Used to identify early signs of mechanical wear or motor imbalance for predictive maintenance [12]

- Optical and vision sensors: Used to detect defects in molded and extruded parts using computer vision-based quality inspection [13].

SENSOR LAYER

Temperature
Process control for
molding and extrusion

Pressure
Injection molding and
hydraulic processes

Vibration
Machinery condition
monitoring

Proximity
Position detection
and automation

Flow
Material control
and dosing

Level
Tanks and silos for
raw materials

Vision
Quality control
and inspection

Humidity
Environmental control
of material

Speed
Motors and
conveyors

Energy
Power consumption
monitory

Force/Torque
Forming process
control

CONTROL ELEMENTS

PLC
Process
automation
and control logic

Microcontroller
Embedded control
and data processing

Controller
Specialized
control
units

HMI
Human Machine
interface

Industrial PC
Advanced processing
and analytics

SCADA/MES
Production
management

Figure 6.3 Sensor layer components, along with control elements, as part of the Industry 4.0 technology architecture. Sensors enable the continuous acquisition of critical process data, while control elements execute automated decisions and intelligent analysis

Smart sensors require the following desirable technical characteristics:

- High sampling rates (10 kHz to 1 MHz) to capture transient energy fluctuations

- Robust communication interfaces (Modbus, OPC-UA, MQTT, BLE, LoRaWAN) for secure data transmission

- Low power consumption for sustainable and long-term deployment

- Embedded edge computing capabilities for local data processing.

Sensorless measurement systems

Sensorless measurement technologies represent a significant advancement in Industry 4.0 solutions for polymer processing, allowing real-time estimation of essential process parameters without requiring direct physical sensors. These techniques use advanced algorithms, machine learning models, and AI-driven analytics to determine values such as torque, viscosity, and melt flow index by assessing indirect process variables. Consequently, sensorless measurement systems offer more reliability, reduced maintenance costs, and enhanced operational efficiency, especially in harsh industrial settings where traditional sensors may fail due to extreme temperatures, high pressures, or contamination [10].

In polymer processing, sensorless measurement technologies have shown potential for enhancing energy efficiency, minimizing operational disruptions, and refining process control [3]. For instance, melt viscosity estimation during extrusion processes can be performed by analyzing motor torque data, removing the requirement for in-line viscometers. Similarly, in injection molding, material flow rates can be accurately predicted without physical flow meters by employing machine learning models that relate injection pressure and screw speed to polymer behavior.

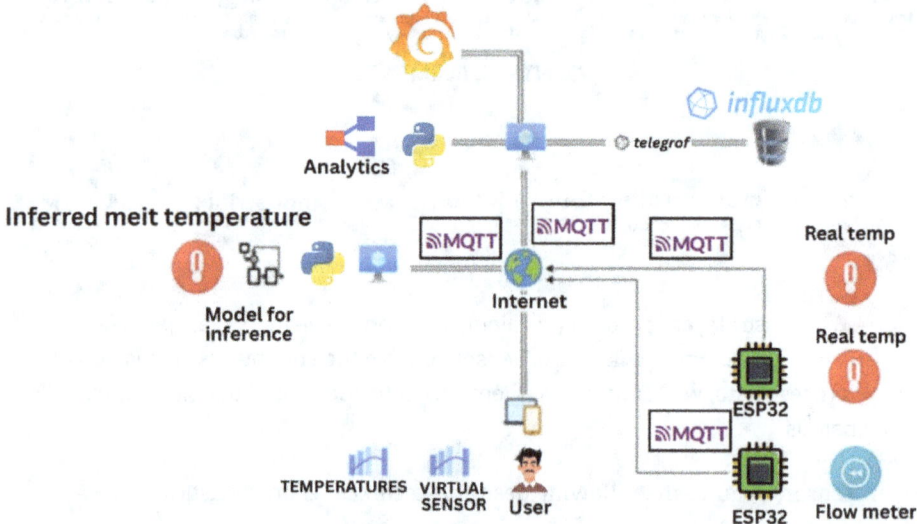

Figure 6.4 Architecture of a sensorless measurement system based on melt temperature inference. The solution integrates IoT nodes (ESP32), MQTT communication, InfluxDB database storage, analytical processing with Python, and real-time visualization

Figure 6.4 illustrates the architecture of a sensorless measurement system enabled by Industry 4.0 technologies for estimating melt temperature in polymer extrusion processes. In this system, ESP32 devices function as nodes that collect data from actual sensors (temperature and flow rate) and transmit it using MQTT protocol. The data is then sent to cloud infrastructure, where it is stored in InfluxDB via the Telegraf agent and visualized on platforms such as Grafana. A Python-based analytics module utilizes the real data to feed a melt temperature inference model with virtual sensors, which estimate this variable without needing a dedicated physical sensor. Finally, the system enables user interaction through web or mobile interfaces, displaying both real and inferred data in real time.

As the examples have shown, implementing sensorless measurement technologies in polymer manufacturing has numerous advantages. One of the most significant benefits is the elimination of sensor failures due to contamination, polymer degradation, or extreme process conditions, which are usual challenges in high-temperature and high-pressure polymer extrusion and molding applications [3]. Furthermore, these technologies significantly lower installation and maintenance costs by eliminating the need for periodic sensor calibration and replacement [3]. By integrating AI-based models and predictive analytics, sensorless measurement techniques enhance process accuracy and enable real-time adjustments, ensuring optimal energy usage and material efficiency. Their ability to seamlessly integrate with IoT and IIoT architectures makes them ideal for modern Industry 4.0-enabled production environments, supporting data-driven decision-making and sustainable manufacturing practices.

Industrial energy meters and actuators

Energy efficiency monitoring in polymer processing relies on industrial energy meters that measure power usage, harmonic distortion, and load imbalances. These meters enable real-time load balancing, energy audits, and anomaly detection in polymer manufacturing plants. The main types of industrial energy meters are:

- Smart power meters: Measure active and reactive power consumption in machines

- Current transformers (CTs): Monitor electrical current variations in motors

- Harmonic analyzers: Detect energy inefficiencies caused by power quality issues.

Industrial actuators are also key components, as they dynamically control heating, cooling, and pressure regulation to optimize process conditions. The main technical requirements for energy meters and actuators are:

- Accuracy class of ±0.2% for high-precision energy monitoring

- Modbus-TCP, OPC-UA, or MQTT integration for seamless data exchange

- Real-time event logging with timestamps for advanced diagnostics.

6.2.1.2 Industrial Communication Protocols

For seamless connectivity, the sensor layer must communicate efficiently with the upper layers of the Industry 4.0 architecture. This function requires robust industrial communication protocols that manage high-speed, secure, and scalable data transmission [11]. The most common communication protocols employed at the industry level are [11]: Modbus RTU/TCP (a traditional wired industrial automation protocol), OPC-UA (which ensures secure, vendor-neutral data exchange between devices), MQTT (Message Queuing Telemetry Transport, a low-bandwidth, lightweight protocol for IIoT applications), and BLE (Bluetooth Low Energy) and LoRaWAN (wireless protocols for battery-operated sensors). Each protocol has strengths depending on latency, bandwidth, and security requirements, making hybrid connectivity models the most efficient in modern polymer processing plants.

6.2.1.3 Challenges and Best Practices

While the sensor layer is essential for improving energy efficiency, its implementation presents several technical and operational challenges that must be addressed to ensure reliable and accurate data acquisition. Issues such as sensor calibration, high data generation rates, and cybersecurity risks can impact the effectiveness of smart manufacturing and energy optimization strategies [9]. Addressing these challenges through proactive sensor management, edge computing, and strong cybersecurity measures can help polymer manufacturers ensure the effective deployment of the sensor layer.

One of the primary challenges in deploying a robust sensor layer is sensor drift and accuracy degradation over time. Sensors that monitor temperature, pressure, power consumption, and material properties may lose accuracy due to prolonged exposure to harsh process conditions, contamination, or aging components [10]. This issue is particularly critical in polymer extrusion and injection molding, where precise temperature and pressure measurements directly impact process stability and energy efficiency. Furthermore, wireless industrial environments introduce interference risks, especially in factories where multiple communication protocols (such as Wi-Fi, Bluetooth, and LoRaWAN) operate simultaneously. Signal degradation and data packet loss can hinder real-time monitoring and process automation, leading to inefficient energy usage and increased operational costs. Another key challenge is data overload, as high-frequency sensor readings generate vast amounts of data, improving storage and processing costs if not managed effectively [11].

Best practices for implementing the sensor layer should be followed to overcome the identified challenges. One of the most effective strategies is regularly calibrating smart sensors and integrating self-diagnostic capabilities to detect drift and automatically adjust readings. This approach ensures data accuracy while minimizing manual maintenance efforts. Additionally, edge processing techniques should be leveraged to filter and preprocess data locally before transmission to central systems. By performing real-time data compression and anomaly detection at the edge, unnecessary data transmission can be reduced, optimizing bandwidth usage and computational resources. Another essential aspect is cybersecurity, as industrial IoT systems are increasingly vulnerable to unauthorized access, data breaches, and cyberattacks. Implementing end-to-end encryption, secure authentication mechanisms, and strict access control policies can protect critical energy consumption and production data from external threats [9].

6.2.2 Communication Layer

The communication layer ensures reliable and secure communication between shop-floor devices, edge servers, and cloud systems. This layer is the backbone of Industry 4.0 solutions in polymer processing, connecting industrial equipment with centralized control systems to enable remote monitoring, automation, and real-time decision-making [14]. The communication layer enables real-time communication between the sensor layer (comprising sensors, actuators, and devices) and the upper layers, which are responsible for data processing, analytics, and enterprise-level decision-making. By establishing seamless data transmission pathways, the communication layer ensures that information flows efficiently across distinct levels of the production environment. In polymer processing, efficient networking is crucial for energy monitoring, predictive maintenance, and process optimization, allowing seamless data exchange across extruders, injection molding machines, robotics, and ERP systems [11, 14].

Several connectivity technologies support industrial communication within this layer. Ethernet/IP and Profinet are commonly used for high-speed wired communication, ensuring real-time industrial automation with minimal delay [11]. These protocols are essential in polymer processing plants where precise equipment synchronization, such as extruders, injection molding machines, and robotic arms, is critical. For wireless connectivity, 5G and LoRaWAN offer robust solutions for remote plant operations. The high bandwidth and low latency of 5G enable real-time data transmission over large distances, making it suitable for applications that require rapid feedback loops, such as energy consumption monitoring and predictive maintenance. Meanwhile, Wi-Fi 6 and Bluetooth Low Energy (BLE) provide power-efficient communication for mobile and handheld devices, ensuring seamless data access for operators and technicians working within the production facility.

A key consideration in network architecture is the balance between edge and cloud communication. Edge computing processes data locally within the factory, reducing reliance on external servers and enabling real-time decision-making with minimal latency [11]. This is particularly beneficial for energy efficiency applications, where immediate responses to fluctuations in power consumption or equipment performance are required [3]. On the other hand, cloud computing plays a crucial role in storing historical production data, supporting long-term analysis, predictive analytics, and process optimization [11]. By leveraging both edge and cloud capabilities, polymer manufacturers can create a hybrid system that maximizes real-time efficiency while enabling advanced data-driven strategies for energy management.

The communication layer offers multiple benefits for energy efficiency and process optimization in polymer processing [6, 11]. First, it reduces latency in mission-critical applications, ensuring that data-driven control mechanisms respond instantly to changes in production parameters. Second, it enhances cybersecurity by incorporating encrypted communication protocols, protecting sensitive production data from cyber threats. Lastly, it enables remote monitoring and predictive maintenance, allowing manufacturers to detect inefficiencies, prevent equipment failures, and optimize energy consumption across multiple facilities.

Future advances in 6G-enabled IIoT, blockchain-secured industrial communication, and AI-driven network management will further enhance the efficiency and resilience of industrial networks in polymer production. To meet the high-performance demands of Industry 4.0, the communication layer must support:

- Low latency, high bandwidth, and real-time communication
- Interoperability across heterogeneous industrial devices
- Robust cybersecurity mechanisms to protect critical manufacturing data
- Scalability for expanding IIoT applications.

This section explores the communication layer's architecture, protocols, and key components in polymer processing plants.

6.2.2.1 Key Components

Wired and wireless industrial networks

Industrial networks provide the physical and logical communication infrastructure to interconnect sensors, controllers, machines, and enterprise systems. Once a device collects data, it must transmit it to the cloud or another network. The choice between wired and wireless networks depends on the application requirements in polymer processing. Modern Industry 4.0 architectures use a hybrid model, combining wired and wireless networks to achieve optimal performance, reliability, and scalability [6].

Wired networks remain the gold standard in industrial automation due to their high reliability, low latency, and robustness in noisy environments [11]. Standard wired protocols include Ethernet/IP (a standard for industrial automation with real-time control capabilities), PROFINET (optimized for high-speed motion control and process automation), and Modbus TCP/IP (a legacy protocol widely used for industrial controllers and energy meters).

Wireless networks provide flexibility and cost savings in industrial environments where wired connections are impractical. Figure 6.5 illustrates various wireless communication protocols that enable data transmission between industrial equipment, edge devices, and cloud computing platforms [6]. Often, the specific connectivity option will depend on the IoT application [11]. However, they must be secure, interference-resistant, and scalable. Typical wireless communication standards include Wi-Fi 6 (IEEE 802.11ax, which supports high-bandwidth data transmission between machines and cloud servers), 5G Industrial Networks (enabling ultra-low latency of 1 ms and high device density for real-time process monitoring), LoRaWAN (Long Range Wide Area Network, ideal for low-power, long-range IIoT sensors in large polymer processing plants), and BLE (Bluetooth Low Energy) & Zigbee (used for short-range communication between sensors and controllers).

Figure 6.5 illustrates a comparative map of wireless communication technologies used in industrial environments, classified according to their range and data rate and power consumption. This visual representation is useful for selecting appropriate connectivity technologies based on the balance between range, data rate, power consumption, and cost, considering the specific requirements of each polymer processing plant or industrial operation [12]. Ethernet and Wi-Fi are commonly used in factories for high-speed connectivity within large production areas. In contrast, Bluetooth is typically used for short-range connections, such as monitoring nearby machinery. Zigbee and LoRa, on the other hand, are well-suited for low-power sensor networks with long-range capabilities, making them ideal for environments where constant power is unavailable or machines are widely spaced apart. LTE and 5G provide high-speed mobile connectivity for real-time transmission of large data volumes to the cloud, enabling greater flexibility in sensor placement and network coverage across large industrial environments.

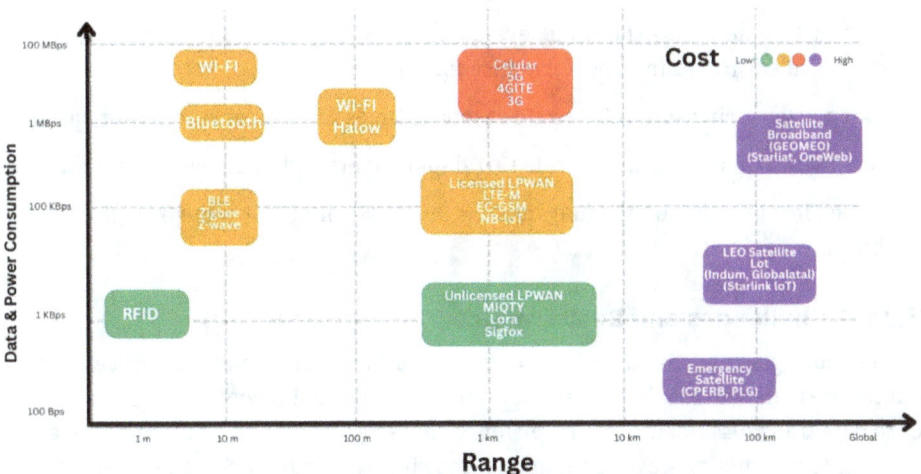

Figure 6.5 Comparison of industrial wireless technologies based on range, transmission speed, and power consumption. Technologies are categorized by operating range (X-axis) and data rate/power consumption (Y-axis), and identified by relative cost tier

Cybersecurity and data integrity mechanisms

With increasing connectivity in polymer processing plants, cybersecurity is a critical challenge. A compromised network can lead to data breaches, production downtime, and energy inefficiencies. Some of the cybersecurity threats in industrial networks are unauthorized access to IIoT devices, "man-in-the-middle" attacks compromising data integrity, and ransomware threats affecting production operations. In this context, some best practices for secure network architecture are [15]:

- Role-based access control (RBAC), which restricts unauthorized access to network devices

- End-to-end encryption, which uses TLS/SSL protocols for secure machine-to-machine communication

- Network segmentation, which separates operational technology (OT) networks from IT networks to mitigate security risks

- Regular firmware updates and AI-driven anomaly detection to help identify cyber threats in real-time.

6.2.2.2 Industrial Communication Protocols

Industrial communication protocols define how data is transmitted and interpreted across Industry 4.0 networks. Beyond the previously mentioned OPC-UA, MQTT and Modbus, we could include PROFINET & EtherCAT, high-speed fieldbus protocols used for motion control in robotic systems [11].

Also, industrial networks should be evaluated based on the following technical characteristics to ensure optimal performance [6, 11]:

- Latency (ensuring sub-millisecond response times for critical automation tasks)

- Bandwidth (managing large-scale IIoT deployments with high data transfer rates)

- Packet loss and reliability (guaranteeing 99.999% uptime for industrial production lines).

6.2.2.3 Challenges and Best Practices

Implementing a robust and scalable communication layer in polymer processing facilities presents several challenges that must be addressed to ensure efficient and secure data transmission. One of the primary challenges in integrating communication layers into polymer processing plants is legacy infrastructure lacking IIoT compatibility [6]. Many factories still operate with traditional wired networks, limiting their ability to support wireless industrial communication and cloud-based analytics [8]. Wireless interference in industrial environments can degrade network performance, especially in facilities with heavy machinery and complex layouts. Another primary concern is cybersecurity, as IIoT devices connected to cloud systems are susceptible to unauthorized access, data breaches, and network attacks, which can compromise sensitive production data and lead to operational disruptions[11].

To optimize the communication layer and ensure seamless data flow, manufacturers should adopt hybrid network architectures that integrate wired and wireless connectivity for improved redundancy and reliability [6, 8, 11]. AI-driven network traffic optimization can help prioritize critical data streams, reducing latency bottlenecks and ensuring that high-priority sensor data reaches processing systems without delay. Strong encryption protocols, firewall protection, and continuous network monitoring are essential to mitigating cybersecurity risks and securing IIoT communication channels. Polymer manufacturers can establish high-performance industrial networks that enhance energy efficiency, process optimization, and predictive analytics in Industry 4.0 environments by addressing these challenges with modern networking strategies.

6.2.3 Processing Layer

The processing layer is responsible for storing, managing, and analyzing industrial data, which is crucial in optimizing energy efficiency in polymer processing. In an Industry 4.0 environment, vast amounts of data are continuously generated from sensors, IIoT devices, and control systems. This layer ensures that energy consumption trends, machine performance metrics, and quality parameters are effectively collected, processed, and transformed into actionable insights for operational improvements [16]. Efficient data management within this layer allows manufacturers to identify inefficiencies, predict maintenance needs, and enhance overall production sustainability.

A fundamental component of this layer is data storage solutions, which determine how industrial data is securely and efficiently maintained. On-premise databases provide localized, secure storage for sensitive production data, ensuring compliance with industry regulations and reducing risks associated with external network vulnerabilities. However, the growing demand for scalability and remote accessibility has led many polymer manufacturers to adopt cloud-based solutions, such as Amazon Web Services (AWS), Microsoft Azure, and Google Cloud. These platforms offer nearly unlimited storage capacity and powerful computing resources, enabling real-time monitoring and advanced data analytics. Alternatively, hybrid storage models combine the best of both worlds, integrating the security and control of on-premise storage with the flexibility and accessibility of cloud infrastructure [16].

Figure 6.6 illustrates these data storage options and their use in industrial manufacturing implementations. On-premise edge computing refers to local data processing at the factory level, where machines, industrial robots, and sensors are connected to an edge gateway. In contrast, edge cloud computing serves as an intermediary infrastructure, aggregating data from multiple factory locations while balancing real-time processing and cloud analytics. At the highest level, public cloud computing involves centralized cloud services, such as Amazon Web Services (AWS), Google Cloud, and Microsoft Azure.

In addition to storage, big data management is crucial for handling the vast volumes of both unstructured and structured data generated during polymer processing [16]. Advanced frameworks provide the computational power needed to analyze large-scale manufacturing datasets, enabling manufacturers to detect patterns in energy usage, optimize machine performance, and improve quality control. Moreover, NoSQL databases are widely used to store sensor-generated data from industrial IoT devices, offering high-speed access and scalability for real-time processing [17].

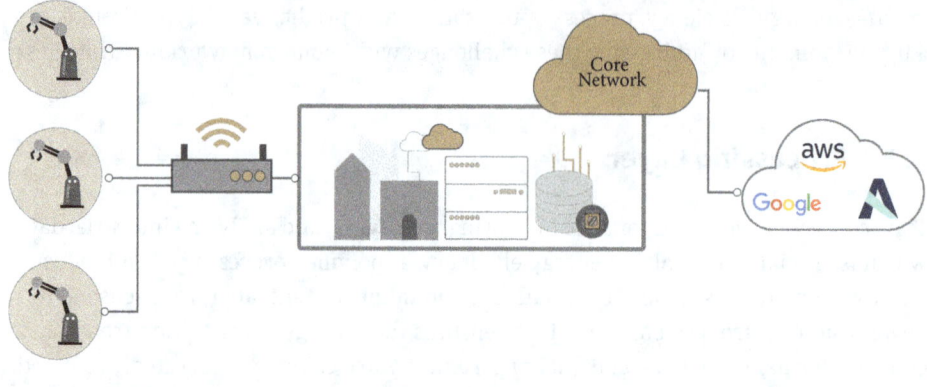

PREMISES EDGE   EDGE CLOUD   PUBLIC CLOUD

Figure 6.6 Hierarchy of data storage models for Industry 4.0 in polymer manufacturing. This includes on-premise edge computing for real-time operations, hybrid edge-cloud architectures for distributed analytics, and public cloud platforms like AWS and Azure for centralized data processing and scalability

The processing layer offers several key benefits for Industry 4.0 implementations in polymer production. First, it enables real-time analytics that help manufacturers make data-driven decisions for improving energy efficiency. Companies can implement optimization strategies that reduce waste and lower operational costs by continuously analyzing energy consumption patterns. Second, it facilitates the integration of machine learning algorithms for predictive maintenance, helping to prevent unexpected equipment failures and extending the lifespan of production machinery. Lastly, it ensures secure, scalable, and efficient data management, providing a robust infrastructure that supports the growing data needs of modern polymer processing plants. As Industry 4.0 technologies evolve, the processing layer remains a critical foundation for achieving higher efficiency, reduced energy consumption, and increased competitiveness in polymer manufacturing.

6.2.3.1 Key Components

In polymer processing, where extruders, injection molding machines, and blow-molding systems operate under energy-intensive conditions, it is imperative to process large volumes of data efficiently. This layer enables real-time data processing, allowing machine learning algorithms and AI models to optimize energy consumption, identify process inefficiencies, and enhance predictive analytics for maintenance strategies [9]. Key responsibilities of this layer include real-time data processing for monitoring and controlling energy consumption, data aggregation and structuring to enable advanced analytics, integration with edge and cloud computing resources for scalable processing, and support for AI-driven models for predictive and prescriptive analytics. The processing layer integrates several technological components that ensure effective data management and real-time analytics. These include:

- Industrial data processing platforms, which manage high-speed data acquisition, filtering, and real-time computation from connected industrial devices

- High-performance computing (HPC) systems, which in polymer plants enable rapid simulation of energy usage models, facilitating real-time optimization

- Data lakes and warehouses, which are storage solutions that allow efficient querying, retrieval, and structuring of large-scale industrial data

- Middleware solutions, which serve as communication bridges between sensor networks, cloud systems, and enterprise software (ERP, MES, SCADA).

6.2.3.2 Industrial Protocols, Software, and Tools

Software and tools for data processing are crucial in modern polymer manufacturing environments. Apache Hadoop & Apache Spark are big data processing frameworks that oversee large-scale manufacturing datasets generated across production lines. NoSQL databases are optimized for unstructured sensor data storage [17], with InfluxDB [18] being particularly valuable in plastic processing contexts due to its time-series database architecture. InfluxDB excels at handling the continuous streams of time-stamped data from temperature sensors, pressure gauges, and energy meters found throughout polymer processing equipment. InfluxDB's query language also facilitates real-time monitoring of critical variables like melt temperature, pressure profiles, and energy consumption patterns, enabling plastic manufacturers to detect deviations that could lead to quality issues or energy waste. Edge analytics platforms (FogHorn, AWS Greengrass, Microsoft Azure IoT Edge) enable on-premise data processing before sending it to the cloud. At the same time, AI/ML Frameworks facilitate predictive modeling for energy efficiency in industrial systems [15].

In the context of plastic manufacturing, time-series databases offer several industry-specific advantages. They allow processors to establish historical baselines for optimal processing windows, tracking how parameters like barrel temperature zones, screw speeds, and melt pressure affect energy consumption and product quality over time. This historical data becomes invaluable for optimizing energy-intensive processes, as it helps identify conditions that minimize energy use while maintaining product specifications. Furthermore, the ability to handle high-cardinality data makes it suitable for facilities with multiple production lines running different polymers and products simultaneously. The database can tag and organize readings by machine ID, polymer type, product SKU, and other metadata, enabling comparative analysis across production scenarios [18]. As plastic manufacturers face increasing pressure to reduce energy consumption and carbon footprints, these time-series databases become essential tools for continuous improvement initiatives, regulatory compliance reporting, and sustainability efforts.

6.2.3.3 Challenges and Best Practices

Implementing the processing layer in Industry 4.0 for polymer processing introduces several challenges that must be carefully managed to ensure efficiency and reliability [15–17]. One of the primary concerns is the high computational and storage costs associated with managing vast amounts of industrial data. The real-time collection and analysis of sensor data from extrusion lines, injection molding machines, and other polymer processing equipment requires significant computing power, which increases operational expenses. Moreover, maintaining data integrity and consistency is critical, as inconsistent sensor readings or network disruptions can compromise the accuracy of predictive models and optimization strategies. Scalability concerns also arise as factories expand their adoption of IoT and IIoT devices, requiring robust infrastructure to oversee increasing data volumes without compromising performance.

Manufacturers should adopt several best practices to optimize the processing layer and overcome these challenges [15, 16]. First, leveraging edge computing for localized data processing is essential to minimize latency and network load by executing real-time calculations closer to the data source. This approach reduces the need for constant cloud communication and enhances the responsiveness of energy efficiency optimization systems. Additionally, implementing scalable cloud storage solutions, particularly hybrid architectures that combine on-premise storage with cloud services, balances security, flexibility, and cost-effectiveness. These solutions allow manufacturers to store high-priority operational data locally while leveraging cloud-based platforms for historical data analysis and predictive modeling.

Another critical strategy is using AI-driven data cleaning algorithms, which improve data accuracy by identifying and correcting anomalies caused by sensor errors, network fluctuations, or inconsistent inputs [17]. High-quality, reliable data is essential for predictive analytics, energy optimization, and adaptive process control in polymer manufacturing. Furthermore, ensuring cybersecurity measures is a top priority, as industrial data contains sensitive operational insights. To mitigate cybersecurity threats, manufacturers should implement end-to-end encryption, firewall protection, and multi-factor authentication, safeguarding their processing layer against cyberattacks and unauthorized access.

6.2.4 Intelligence Layer

The intelligence layer represents the core of advanced decision-making in Industry 4.0 solutions for energy efficiency in polymer processing. By leveraging machine learning (ML) and AI, this layer enables real-time optimization of energy consumption, predictive maintenance, and quality control, ensuring that manufacturing processes are more sustainable and cost-effective [13]. The intelligence layer is built upon sophisticated algorithms that analyze large datasets, extract meaningful patterns, and generate action-able insights for continuous process improvement.

A key component of this layer is AI-driven process optimization, which uses advanced computational models to fine-tune production settings in polymer processing. Neural networks are critical in improving energy efficiency, as they can predict the optimal machine configurations to minimize power consumption while maintaining produc-tion quality [10]. Additionally, computer vision techniques based on convolutional neu-ral networks (CNNs) are employed for real-time defect detection in molded and ex-truded polymer products, significantly reducing material waste and improving product consistency [13]. Another transformative application is digital twins, which create vir-tual simulations of production environments [18]. These digital representations allow manufacturers to evaluate various operating scenarios, optimize energy usage, and re-fine process parameters before implementing changes in the actual production line.

In addition to process optimization, predictive maintenance is another critical func-tionality within the intelligence layer [12]. Machine learning models, such as K-nearest neighbors (KNN), random forest, and long short-term memory (LSTM), analyze sensor data from industrial machinery to forecast potential failures before they occur. This capability is essential in reducing unplanned downtime, as early fault detection en-ables timely interventions, preventing costly disruptions in production [12]. Moreover, automated fault diagnosis systems powered by AI-driven analytics have been shown to reduce unexpected breakdowns by up to 25%, further enhancing operational reliability and energy efficiency.

The benefits of integrating AI and ML technologies into polymer processing are sub-stantial. First, these technologies enhance production efficiency by enabling adaptive process control, allowing machinery to adjust dynamically based on real-time condi-tions. Second, AI-driven energy optimization reduces power consumption by ensuring that machines operate within the most efficient parameters, directly lowering produc-tion costs. Finally, real-time defect detection minimizes material waste, ensuring that only high-quality products reach the market, thus improving sustainability and profit-ability.

6.2.4.1 Key Components

The key functions of this layer include AI-driven optimization of polymer processing parameters (e.g., temperature, pressure, and extrusion speed), machine-learning-based predictive maintenance, minimizing unexpected downtimes, digital twin technology for virtual process simulation and energy optimization, and anomaly detection and fault diagnostics to reduce energy and material wastage [19]. Figure 6.7 illustrates the six-step process for leveraging AI and data analytics in polymer processing. AI is instrumental in optimizing energy consumption by predicting ideal process parameters, reducing down-time through predictive maintenance of polymer processing equipment, improving production quality using machine learning and computer vision techniques, and enhancing process automation, leading to more efficient and sustainable operations. This layer serves as the decision-making hub, continuously learning from historical and real-time data to suggest improvements in efficiency and productivity.

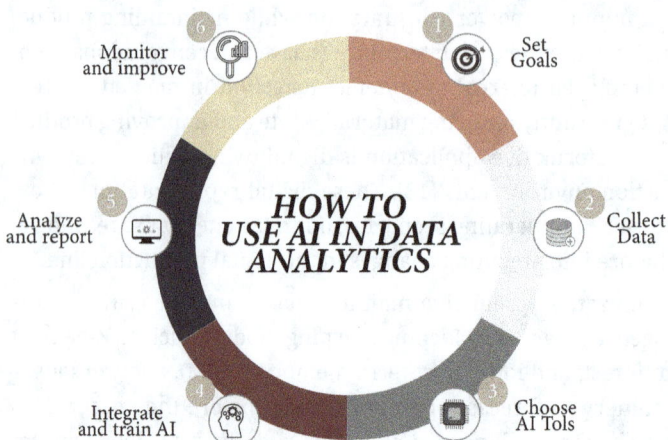

Figure 6.7 Workflow for implementing AI-driven data analytics in the intelligence layer of Industry 4.0 architectures, aimed at enhancing energy efficiency in polymer processing by supporting predictive decision-making and operational optimization

AI and machine learning models

Some of the models used in this context include supervised and unsupervised learning algorithms, which are employed to identify patterns in energy consumption and predict equipment failures in polymer processing plants. These algorithms can analyze historical data to establish correlations between processing conditions and energy efficiency outcomes, helping operators optimize production parameters. Deep learning (DL) Networks, particularly convolutional neural networks (CNNs), are applied for quality control and energy-efficient defect detection in extruded polymers and injection-molded parts [10]. By identifying defects early in the production process, these networks help prevent energy waste associated with producing non-conforming products.

Additionally, recurrent neural networks (RNNs) and long short-term memory (LSTM) models are applied in time-series analysis for energy forecasting [12] in polymer manufacturing facilities. These advanced neural network architectures excel at processing sequential data, making them ideal for predicting energy consumption patterns based on production schedules, material properties, and environmental factors, thereby enabling more precise energy management strategies in plastic processing operations.

Digital twin technology

Digital twins represent virtual models replicating real-world polymer processing systems to optimize energy efficiency in manufacturing environments. These sophisticated digital replicas capture the physical, thermal, and mechanical characteristics of polymer processing equipment such as extruders, injection molding machines, and thermoforming systems. Digital twins simulate different production scenarios, allowing engineers to identify energy-saving measures before real-world implementation [18]. This is particularly valuable in polymer processing, where energy costs represent 15–20% of operational expenses. By manipulating variables like barrel temperature profiles, screw speeds, cooling rates, and material formulations in the virtual environment, engineers can determine optimal processing parameters that minimize energy consumption while maintaining product quality. These simulations also enable manufacturers to evaluate the energy impact of different polymer grades, additives, and recycled content without disrupting production flows. Furthermore, digital twins can model the entire polymer processing plant as an integrated system, identifying opportunities for waste heat recovery, load balancing, and peak demand management across multiple machines and processes, leading to comprehensive energy optimization strategies that would be difficult to develop through conventional methods.

Edge AI for real-time energy analytics

AI processing at the edge reduces latency and enables faster response times for energy management in polymer processing operations. This solution ensures continuous process optimization without reliance on cloud-based computations, which are critical in high-speed plastic manufacturing where milliseconds matter. Edge AI devices deployed directly on extrusion lines, injection molding machines, and thermoforming equipment can instantly analyze sensor data and make real-time adjustments to process parameters, optimizing energy consumption within the same processing cycle [18]. This capability is significant for energy-intensive processes like polymer melt temperature control, where slight deviations can significantly impact energy efficiency. These edge systems can also function independently during network outages by processing data locally, ensuring uninterrupted production and energy management. Additionally, edge AI reduces bandwidth requirements and associated energy costs by filtering and processing data locally, sending only relevant insights to central systems rather than raw data streams from hundreds of sensors throughout the plastic processing facility.

Edge AI platforms are particularly valuable in polymer manufacturing contexts where the harsh industrial environment may compromise network reliability. For example, an edge AI system might immediately detect a deviation in heating zone efficiency in an extruder and adjust parameters to compensate, preventing energy waste before it impacts production costs. Meanwhile, data processing frameworks provide the distributed computing power necessary to analyze terabytes of production and energy data across multiple polymer processing lines, identifying facility-wide optimization opportunities that would be invisible when examining individual machines in isolation [17].

6.2.4.2 Industrial Protocols, Software, and Tools

The intelligence layer relies on specialized AI and analytics platforms for real-time decision-making. TensorFlow [20], PyTorch [21], and Scikit-Learn [22] are common AI/ML frameworks for predictive energy modeling in polymer processing operations. Hadoop & Apache Spark manage large-scale industrial datasets for energy analytics across manufacturing facilities. Edge AI platforms (NVIDIA Jetson, Intel OpenVINO, AWS Greengrass) enable on-premise, real-time AI computation directly at production lines.

These software tools form the backbone of intelligent energy management in polymer processing by transforming raw sensor data into operational insights. TensorFlow [20], developed by Google, offers a comprehensive ecosystem for building and deploying machine learning models to analyze complex energy consumption patterns in polymer processing. TensorFlow's distributed computing capabilities enable parallel processing of sensor data from multiple machines, making it ideal for plant-wide energy optimization in large polymer facilities.

PyTorch [21] provides a more dynamic and intuitive approach to neural network development that many polymer process engineers find accessible. Its define-by-run paradigm allows easier debugging and experimentation when developing energy optimization models. PyTorch has proven particularly valuable in injection molding applications for creating models that simultaneously optimize energy consumption across multiple machine parameters such as barrel temperature profiles, injection speed, and cooling time [13]. The framework's strong support for GPU acceleration enables real-time training on production data, allowing models to continuously adapt to changing material characteristics or tooling configurations.

Scikit-Learn [22], while less specialized for deep learning than TensorFlow or PyTorch, offers a comprehensive suite of traditional machine learning algorithms that are often more appropriate for polymer processing applications where data volumes may be limited or interpretability is crucial. Its random forest algorithms and gradient boosting techniques can effectively identify the most energy-intensive stages of polymer processing and help prioritize optimization efforts. In extrusion operations, Scikit-Learn's regression models help establish clear relationships between operational parameters like screw design, polymer viscosity, and energy efficiency, providing actionable insights that process engineers can implement.

6.2.4.3 Challenges and Best Practices

Deployment of the intelligence layer in polymer processing plants presents several challenges that must be addressed to ensure efficient and reliable energy optimization. One of the primary obstacles is the high computational demand required for training and running AI models [11]. Advanced neural networks, predictive analytics, and real-time process optimization require high-performance computing resources, which can be costly and complex. Additionally, data quality issues present a significant challenge [17]. AI algorithms depend on accurate, complete, and high-resolution industrial data, but sensor drift, missing data points, or reading inconsistencies can significantly reduce the accuracy of AI-driven predictions.

Another major challenge in implementing AI-powered intelligence is integration complexity [11]. Polymer processing plants typically operate with various legacy systems, including ERP (enterprise resource planning), MES (manufacturing execution systems), and SCADA (supervisory control and data acquisition). Ensuring seamless interoperability between these systems and AI-driven solutions requires robust APIs, standardized communication protocols, and scalable data processing frameworks. Without proper integration strategies, the Intelligence Layer may fail to deliver actionable insights in real-time production environments.

To overcome these challenges, manufacturers should adopt best practices for deploying AI-powered intelligence in polymer processing. One effective strategy is to deploy AI-powered process control systems that dynamically adjust energy-intensive operations, such as extrusion temperature settings, injection molding cycle times, and cooling system efficiency [3]. These adaptive control strategies enable real-time energy consumption optimization, reducing waste and improving overall efficiency. Also, a hybrid AI architecture combining cloud-based model training with edge computing for real-time deployment is recommended. Cloud platforms, such as AWS, Microsoft Azure, or Google Cloud AI, provide scalable infrastructure for training complex AI models. At the same time, edge deployment ensures that AI-driven optimizations can be executed locally within the plant, minimizing latency and enhancing real-time decision-making.

Finally, self-learning AI models that continuously adapt to process variations offer a significant advantage. These models use reinforcement learning and adaptive algorithms to detect patterns in machine performance, material behavior, and environmental conditions, automatically adjusting operational parameters to maximize energy efficiency. This autonomous adaptation reduces operator dependency, ensuring consistent improvements in energy management.

6.2.5 Application Layer

The application layer is the interface between Industry 4.0 technologies and enterprise management systems, facilitating seamless integration, automation, and optimization of production processes [23]. This layer enables efficient communication and coordination between industrial equipment, real-time monitoring systems, and business management software, focusing on a data-driven approach to decision-making. A fundamental aspect of this layer is integrating with enterprise systems, ensuring that manufacturing operations align with broader business goals. ERP systems, such as SAP, Oracle, and Microsoft

Dynamics, provide centralized inventory management, supply chains, and resource allocation, ensuring that energy-intensive polymer production is conducted optimally. MES bridges the gap between shop-floor operations and business-level decision-making, synchronizing real-time sensor data with production workflows to enhance operational responsiveness. Additionally, SCADA systems allow for real-time supervision and control of industrial equipment, facilitating precise adjustments to optimize energy consumption and reduce production inefficiencies.

The application layer relies on standardized interoperability frameworks for effective data exchange, ensuring seamless connectivity between industrial and business applications [23]. RESTful APIs (application programming interfaces) and WebSocket allow real-time data sharing between Industry 4.0 platforms and ERP systems, enabling automated decision-making based on live production data.

The benefits of the application layer in Industry 4.0 for polymer processing are significant. By automating decision-making, this layer improves operational efficiency, reducing human intervention and minimizing errors. Integrating IoT-driven data with ERP and MES systems enhances traceability and compliance, allowing manufacturers to meet stringent industry regulations while maintaining transparency across production cycles. Most importantly, optimized energy use across multiple manufacturing plants becomes feasible, as centralized monitoring and AI-driven analytics enable precise adjustments in power consumption, material flow, and process parameters.

6.2.5.1 Key Components

The application layer acts as the bridge between production-level data and enterprise decision-making. It consolidates sensor data, operational performance metrics, and predictive insights to enable energy-efficient process optimization. Key functions of this layer include [3, 23]:

- Energy performance monitoring through integrated dashboards displays real-time energy usage and efficiency trends

- AI-driven optimization of processing parameters to reduce energy waste in polymer extrusion, injection molding, and blow-molding operations

- Automated production scheduling based on energy availability and cost considerations

- Integration with business intelligence (BI) systems for strategic energy efficiency planning.

This layer facilitates the seamless interaction between cyber-physical production systems (CPPS) and enterprise-level management, allowing polymer manufacturers to improve productivity while minimizing energy consumption. The application layer consists of several core components that support data visualization, process control, and enterprise-wide decision-making.

Energy management software

- MES provide real-time tracking of production energy efficiency, integrating with shop-floor sensors and industrial controllers [9]

- ERP systems such as SAP, Oracle, and Microsoft Dynamics optimize energy consumption, material flow, and cost management in polymer processing [11]

- SCADA systems enable real-time monitoring and control of polymer manufacturing equipment, ensuring optimal energy efficiency [14].

Human–machine interface (HMI) and dashboarding tools

Industrial HMIs provide operators with real-time insights into energy consumption patterns and process efficiency. Data visualization platforms transform large-scale energy data into actionable insights for plant managers and executives [9]. These visualization tools are critical in making complex energy data accessible and actionable across all levels of polymer manufacturing organizations. Industrial HMIs are the primary interface between operators and processing equipment, displaying real-time energy metrics directly on the factory floor. Modern HMIs in polymer processing plants have evolved beyond basic control panels to become sophisticated dashboards that visualize energy consumption relative to production throughput, material characteristics, and quality metrics. These interfaces enable machine operators to identify energy-intensive processing stages and immediately adjust parameters like barrel temperature profiles, cooling systems, or motor speeds to optimize efficiency while maintaining product specifications.

Figure 6.8 shows a dashboard-style user interface for real-time energy monitoring of an industrial printing line (referred to as "PRINTER 2"), exemplifying a practical application of HMI tools for operational efficiency in Industry 4.0 environments. A time-series graph visualizes the behavior of energy consumption (in watts) over time for three phases identified as I2f3, I2f2, and I2f1, as well as the TOTAL curve (in orange). Also, four gauge-style visual indicators are included, representing the instantaneous power consumption for each phase and the combined total. This type of HMI display offers quick and intuitive feedback to operators or energy managers, facilitating an immediate response to critical situations such as overloads or phase imbalances [9]. The indicators are color-coded (green to red) to indicate proximity to the recommended consumption threshold, helping to maintain efficient control of energy use.

At the management level, platforms like Tableau [24] and Microsoft Power BI [25] aggregate data from multiple production lines and processing technologies to provide a comprehensive view of energy utilization across the facility. Tableau's interactive visualizations allow production managers to drill down from plant-wide energy consumption to individual machine performance, identifying equipment operating outside optimal energy efficiency ranges. The platform's ability to create custom dashboards enables different stakeholders to focus on relevant metrics – maintenance teams can monitor energy anomalies that might indicate equipment issues. At the same time, sustainability officers can track progress toward energy reduction targets.

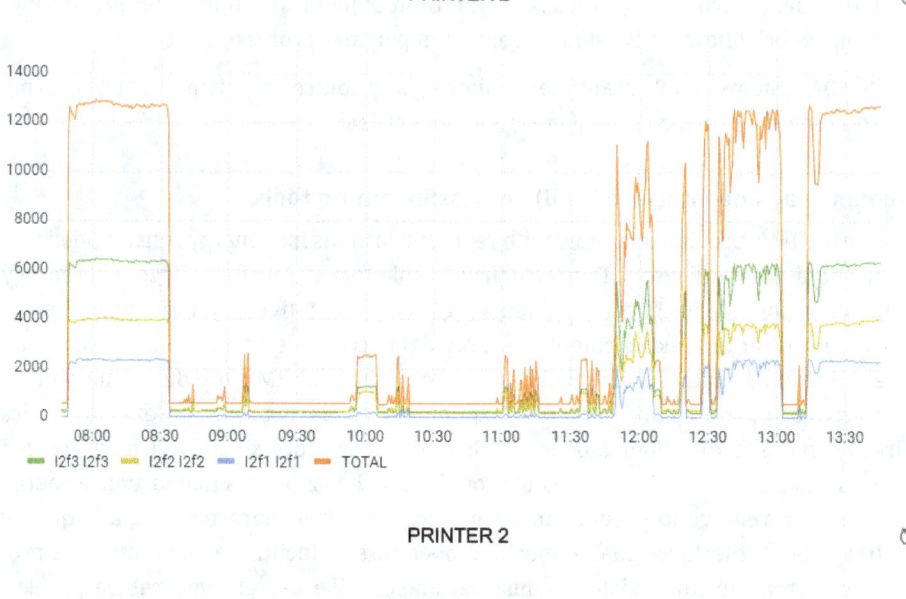

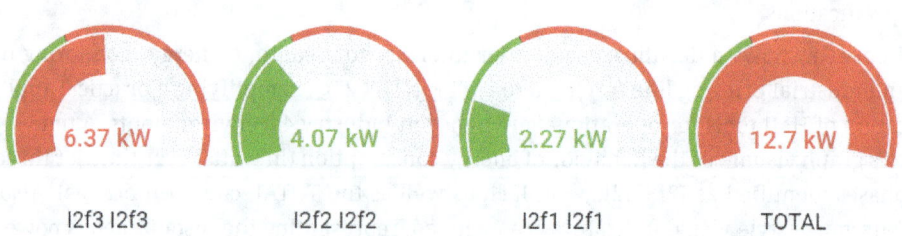

Figure 6.8 Example of a real-time energy monitoring panel for an industrial printing line (IMPRESORA 2), using HMI tools (designed for the case of study). The time evolution of consumption by phase (I2f3, I2f2, I2f1) and the combined total are displayed, along with color-coded instantaneous consumption indicators

With its strong integration with Microsoft's ecosystem, Power BI facilitates connecting energy data with enterprise resource planning systems. It allows correlation between energy consumption and business metrics like production costs, product margins, and carbon footprint calculations. Its advanced analytics capabilities enable predictive energy usage modeling based on production schedules and material types, helping polymer processors optimize electricity procurement and demand management.

The platform's natural language query features make energy data more accessible to non-technical stakeholders. This enables executives to ask plain-language questions about energy performance and receive visual answers that drive strategic decision-making around sustainability initiatives and capital investments in more energy-efficient processing equipment.

Predictive and prescriptive analytics

- AI-powered anomaly detection: Identifies abnormal energy consumption patterns and alerts operators to potential inefficiencies [12]

- Automated process adjustments: Machine learning algorithms optimize polymer processing parameters dynamically.

6.2.5.2 Industrial Protocols, Software, and Tools

To facilitate seamless integration of real-time energy monitoring with enterprise systems, the application layer utilizes various industrial protocols and software solutions. Beyond the already mentioned ERP, MES, SCADA, and HMI, other software platforms that could be involved with this layer are [23]:

- EMS (Energy Management System), which is crucial in monitoring and optimizing energy consumption and reducing the carbon footprint.

- LIMS (laboratory information & management system), which ensures quality control by tracking polymer material properties, reducing defective product batches

- CMMS (Computerized Maintenance Management System), which enhances predictive maintenance to prevent energy losses due to inefficient equipment

- PLC (programmable logic controller), which implements real-time process control and automation, ensuring optimized operations with minimal energy waste.

6.2.5.3 Challenges and Best Practices

The implementation of the application layer in polymer processing plants presents several challenges that must be addressed to ensure effective integration and long-term scalability. One of the primary challenges is legacy system integration. Many polymer manufacturers rely on older MES and ERP platforms that are not designed to support real-time energy analytics or predictive optimization [23]. These legacy systems often lack interoperability with modern SCADA platforms, limiting the ability to synchronize production data with energy consumption insights. As a result, manufacturers may struggle to fully leverage Industry 4.0 applications without significant upgrades or middleware solutions.

Another major obstacle is scalability. Implementing Industry 4.0 applications across multiple plant locations or integrating them with varied production lines can be costly and complex [23]. Each facility may have unique process requirements, equipment configurations, and data infrastructures, requiring customized deployments rather than a one-size-fits-all approach. Manufacturers may experience delays, inefficiencies, and higher implementation costs without a standardized deployment framework. Additionally, data security concerns present a significant challenge. As cloud-based industrial applications become more prevalent, protecting sensitive operational data from cyber threats is critical. Unauthorized access, data breaches, and industrial espionage pose serious risks to manufacturers, particularly when integrating cloud-based Industry 4.0 platforms with existing on-premise systems. Companies may face operational disruptions, compliance issues, and financial losses without robust cybersecurity measures.

To maximize the benefits of the application layer and ensure its successful deployment, manufacturers should adopt best practices tailored to energy-efficient polymer processing [23]. First, companies should adopt modular and scalable software solutions that support future-proof industrial applications. Flexible, cloud-based platforms enable gradual implementation without disrupting existing production processes, ensuring a smooth transition to Industry 4.0. Another crucial best practice is the implementation of AI-driven process automation to optimize energy consumption in real time. Additionally, seamless integration with MES, ERP, and SCADA systems is essential to create a unified energy management framework. Finally, deploying cybersecurity best practices is imperative to protect industrial applications from cyber threats. Implementing multi-factor authentication, encrypted communication channels, and role-based access control (RBAC) enhances data security and prevents unauthorized access. Regular security audits and compliance with international cybersecurity standards further reinforce the resilience of Industry 4.0 applications.

6.3 Implementation of Industry 4.0 Technologies

Adopting Industry 4.0 technologies in polymer processing enhances energy efficiency, operational productivity, and product quality. However, transitioning to a digitalized environment requires a structured implementation roadmap to ensure the successful deployment of these strategies. This section outlines a step-by-step framework for companies integrating Industry 4.0 technologies in polymer manufacturing.

To begin the digital transformation journey, companies must assess their technological, organizational, and operational readiness [11]. A comprehensive Industry 4.0 maturity assessment includes evaluating the current state of processes, energy consumption, and production inefficiencies [9], analyzing the availability of IoT devices and computational resources [14], and identifying workforce skill gaps to establish adequate training programs. A well-defined strategy with clear objectives and key performance indicators (KPIs) is essential for Industry 4.0 implementation. Companies should set energy efficiency targets, such as reducing specific energy consumption by

15% [13], defining predictive maintenance goals to minimize downtime [12], and optimizing production to reduce material waste and improve cycle times [10].

The next step involves selecting and integrating Industry 4.0 technologies for polymer processing. These include IoT and IIoT-based real-time monitoring of extrusion, injection molding, and polymerization processes [19], the use of big data analytics for resource optimization [16], and AI-driven predictive maintenance to anticipate equipment failures [15]. Additionally, smart process automation facilitates AI-based quality control for defect detection and process optimization [9], while cloud and edge computing ensure secure data storage and real-time decision-making [19].

Before full-scale deployment, pilot projects allow companies to validate the benefits of Industry 4.0 technologies by focusing on high-energy consumption processes such as extrusion and injection molding [11]. This phase includes implementing sensor networks for energy monitoring [14] and data-driven optimization to refine process parameters. Once pilot projects yield positive results, companies can scale Industry 4.0 solutions across all production units. Full-scale deployment involves the expansion of IoT, AI, and analytics platforms, complemented by workforce training to ensure effective interaction with modern technologies. Establishing a continuous improvement cycle through real-time analytics dashboards allows for ongoing performance monitoring and decision-making [6].

Measuring the return on investment (ROI) helps evaluate the financial and operational impact of Industry 4.0 implementation. This includes energy savings analysis to assess reductions in specific energy consumption (SEC) [13], productivity improvements through increased throughput and defect reduction, and long-term cost savings from predictive maintenance [15]. Despite its numerous advantages, adopting Industry 4.0 poses challenges, such as data security risks, high initial investments, and technology integration complexities [11]. Emerging trends, including digital twins, decentralized AI models, and blockchain-based supply chain transparency, will further shape the future of polymer manufacturing.

6.4 Case Study

Implementing digital enabling technologies such as the IoT, AI, and cloud computing significantly transform industrial production methods. The increasing digitization of industrial processes allows real-time monitoring of machine performance, energy consumption, and operational efficiency, which leads to informed decision-making and resource optimization [26]. This case study presents the implementation of an Industry 4.0 solution in a plastic manufacturing company, integrating IoT, cloud computing, and AI-based analytics. The objective was to enhance energy efficiency and optimize production processes in key areas such as extrusion, weaving, and printing. The methodol-

ogy involved real-time data acquisition, advanced analytics, and predictive maintenance techniques to identify inefficiencies and propose corrective measures.

System description

The implemented system consisted of a network of IoT sensors deployed in different
machines across the production floor. These sensors measured critical parameters
such as voltage, current, and power consumption, allowing the calculation of real-time
energy usage. The data collected was transmitted via WiFi-enabled microcontrollers to
a cloud-based database (InfluxDB), which was integrated with visualization tools such
as Grafana and Streamlit for monitoring and analysis.

The architecture of the Industry 4.0 solution, shown in Figure 6.9, includes:

- IoT sensors and microcontrollers: Deployed on production machines to measure
  energy consumption and operational parameters

- Cloud-based data storage: A centralized database (InfluxDB) for data collection and
  historical storage

- Data analytics and visualization: The system included a web-based dashboard allowing real-time monitoring and analytics

- Integration with manufacturing execution systems (MES): This action facilitates correlation of energy consumption with production output to assess efficiency.

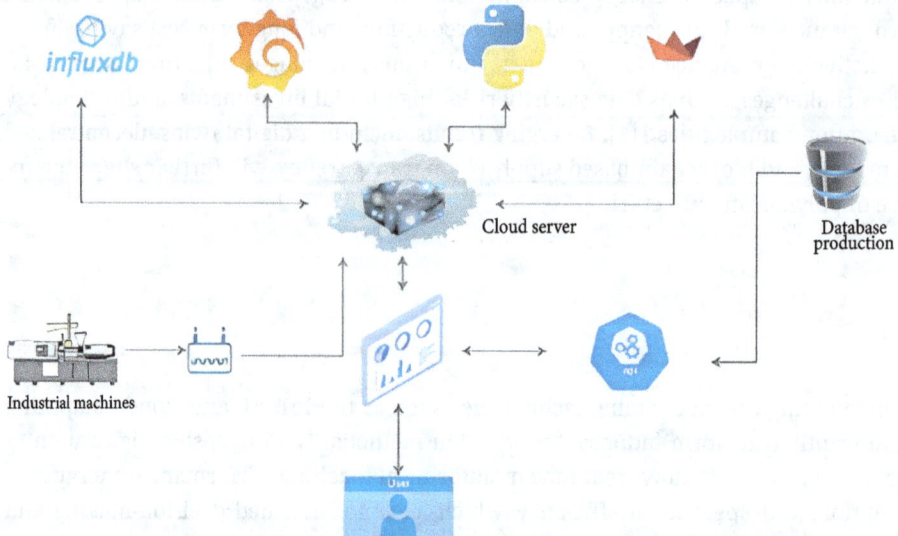

Figure 6.9 System architecture of the Industry 4.0 implementation for energy monitoring and optimization in polymer processing. It illustrates the integration of IoT sensors, cloud-based databases (InfluxDB), and analytics dashboards (Grafana and
Streamlit) for real-time visualization and process control

Energy consumption analysis

The energy consumption monitoring system enabled a detailed analysis of machine power profiles. Figure 6.10 presents a real-time monitoring dashboard showing the power profiles of different machines, including looms and printing equipment. The system provided insights into energy usage trends, enabling the identification of anomalous patterns and inefficiencies.

Key findings included:

- Identifying peak energy consumption periods allows adjustments in production scheduling to reduce peak loads

- Detection of abnormal power consumption to indicate potential maintenance needs or suboptimal machine performance

- Comparison of energy usage across different machine types, facilitating the identification of energy-intensive processes that require optimization.

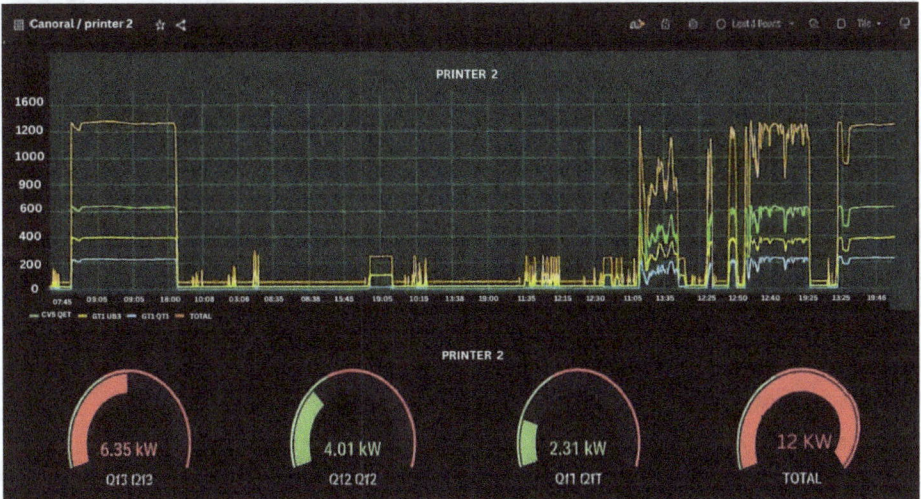

Figure 6.10 Real-time energy monitoring dashboard developed with Grafana for a polymer processing machine (IMPRESORA 2). The interface displays power consumption trends over time, current energy usage per sensor node, and the total system load, enabling operators to detect inefficiencies and optimize machine performance

Production analysis and integration with MES

The platform was integrated with the company's MES, combining production data with energy consumption metrics. This integration enabled the calculation of KPIs such as energy consumption per product unit and overall equipment efficiency (OEE).

Figure 6.11 shows an example of production data visualization, where the system provided real-time insights into material processed, machine efficiency, and productivity trends. The analysis of production data helped with the following issues:

- Identifying production bottlenecks: Allowing workload redistribution to optimize efficiency

- Assessing machine-specific energy consumption: Enabling targeted improvements in machine settings

- Improving scheduling and resource allocation: Reducing idle times and energy waste.

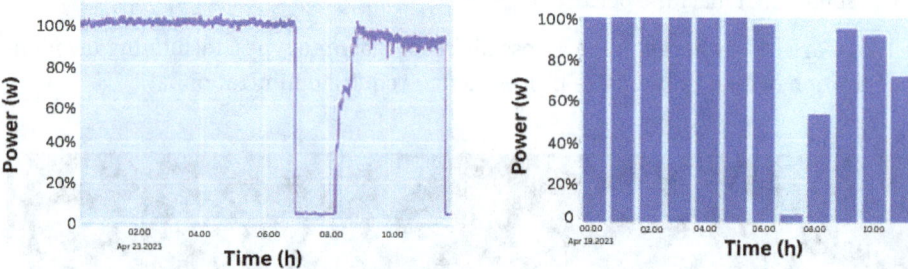

Figure 6.11 Energy consumption profile of Extruder 1 on 25 April 2023. The left graph shows the total instantaneous power [W] over the monitored period, highlighting operational patterns and idle phases. The right chart presents the average hourly power consumption, facilitating a comparative energy demand analysis throughout the production cycle

Results and key benefits

The implementation of the Industry 4.0 solution provided significant improvements in energy efficiency and production performance. The main results included:

- Reduced energy consumption by 15% in extrusion and injection molding processes through optimized machine operation settings

- Decrease in unplanned downtime by 20% due to predictive maintenance strategies

- Improvement in defect detection accuracy to 95% using AI-powered quality control systems.

Integrating IoT, cloud computing, and AI in polymer manufacturing has proven to be a valuable approach to enhancing energy efficiency and operational productivity. This case study highlights the importance of real-time monitoring, data analytics, and predictive maintenance in optimizing industrial processes. Future work should focus on expanding the system's capabilities, incorporating edge computing for faster data processing, and leveraging advanced AI algorithms for further optimization.

References

[1] K. Schwab, "The fourth industrial revolution", World Economic Forum, 2016

[2] B. C. Menezes, J. D. Kelly, A. G. Leal, "Identification and design of Industry 4.0 opportunities in manufacturing: Examples from mature industries to laboratory level systems", *IFAC-PapersOnLine*, 2019, vol. 52, pp. 2494–2500, DOI: 10.1016/j.ifacol.2019.11.581

[3] C. Abeykoon, A. McMillan, B. K. Nguyen, "Energy efficiency in extrusion-related polymer processing: A review of state of the art and potential efficiency improvements", *Renewable and Sustainable Energy Reviews*, 2021, vol. 147, article no. 111219, DOI: 10.1016/j.rser.2021.111219

[4] J. M. Müller, D. Kiel, K.-I. Voigt, "What drives the implementation of Industry 4.0? The role of opportunities and challenges in the context of sustainability", *Sustainability*, 2018, vol. 10, p. 247, DOI: 10.3390/su10010247

[5] G. Arana-Landín, N. Uriarte-Gallastegi, B. Landeta-Manzano, I. Laskurain-Iturbe, "The contribution of lean management – Industry 4.0 technologies to improving energy efficiency", *Energies*, 2023, vol. 16, p. 2124, DOI: 10.3390/en16052124

[6] F. Wortmann, K. Flüchter, "Internet of Things: Technology and value added", *Business & Information Systems Engineering*, 2015, vol. 57, pp. 221–224, DOI: 10.1007/s12599-015-0383-3

[7] K. Zhou, Taigang Liu, and Lifeng Zhou, "Industry 4.0: Towards future industrial opportunities and challenges", in: 12th International Conference on Fuzzy Systems and Knowledge Discovery (FSKD), IEEE, 2015, pp. 2147–2152, DOI: 10.1109/FSKD.2015.7382284

[8] H. Kagermann, J. Helbig, W. Wahlster, A. Hellinger, "Recommendations for implementing the strategic initiative INDUSTRIE 4.0: Securing the future of German manufacturing industry" [Final Report of the Industry 4.0 Working Group], Forschungsunion, 2013, *https://books.google.com.co/books?id=AsfOoAEACAAJ*

[9] T. Schuett et al., "Application of digital methods in polymer science and engineering", *Advanced Functional Materials*, 2024, vol. 34, article no. 2309844, DOI: 10.1002/adfm.202309844

[10] C. Abeykoon, "Design and applications of soft sensors in polymer processing: A review", *IEEE Sensors Journal*, 2019, vol. 19, pp. 2801–2813, DOI: 10.1109/JSEN.2018.2885609

[11] P. Gupta et al., "Industrial internet of things in intelligent manufacturing: A review, approaches, opportunities, open challenges, and future directions", *International Journal on Interactive Design and Manufacturing (IJIDeM)*, 2022, DOI: 10.1007/s12008-022-01075-w

[12] R. Kumari, K. Saini, A. Anand, "Predictive analytics to improve the quality of polymer component manufacturing", *Measurement: Sensors*, 2022, vol. 24, article no. 100428, DOI: 10.1016/j.measen.2022.100428

[13] M. El Ghadoui, A. Mouchtachi, R. Majdoul, "Exploring and optimizing deep neural networks for precision defect detection system in injection molding process", *Journal of Intelligent Manufacturing*, 2024, vol. 36, pp. 2897–2914, DOI: 10.1007/s10845-024-02394-3

[14] Z.-H. Wang, Y.-T. Li, Y.-C. Wu, "Design of intelligent manufacturing IoT sensing system for polymer process monitoring", *The International Journal of Advanced Manufacturing Technology*, 2023, vol. 129, pp. 2933–2947, DOI: 10.1007/s00170-023-12510-x

[15] S.-C. Oh, A. J. Hildreth, *Analytics for Smart Energy Management: Tools and Applications for Sustainable Manufacturing* (1st edition), Springer, 2016, DOI: 10.1007/978-3-319-32729-7

[16] A. Gandomi, M. Haider, "Beyond the hype: Big data concepts, methods, and analytics", *International Journal of Information Management*, 2015, vol. 35, pp. 137–144, DOI: 10.1016/j.ijinfomgt.2014.10.007

[17] I. H. Sarker, "Data science and analytics: An overview from data-driven smart computing, decision-making and applications perspective", *SN Computer Science*, 2021, vol. 2, p. 377, DOI: 10.1007/s42979-021-00765-8

[18] Y. S. Kang, J. G. Choi, S. M. Yang, S. Lim, D. C. Kim, H. W. Park, "Artificial intelligence augmented digital twin for improving the machinability in a robotic carbon fiber reinforced plastics machining process", *J. Intell. Manuf.*, jun. 2025, doi: 10.1007/s10845-025-02642-0.

[19] L. Waltersmann, S. Kiemel, J. Stuhlsatz, A. Sauer, R. Miehe, "Artificial intelligence applications for increasing resource efficiency in manufacturing companies – A comprehensive review", *Sustainability*, 2021, vol. 13, p. 6689, DOI: 10.3390/su13126689

[20] Martín Abadi et al., "TensorFlow: Large-Scale Machine Learning on Heterogeneous Systems". 2015. *https://www.tensorflow.org/*

[21] A. Paszke et al., "PyTorch: An Imperative Style, High-Performance Deep Learning Library", en *Advances in Neural Information Processing Systems 32*, Curran Associates, Inc., 2019, pp. 8024–8035. [En línea]. Disponible en: http://papers.neurips.cc/paper/9015-pytorch-an-imperative-style-high-performance-deep-learning-library.pdf

[22] F. Pedregosa et al., "Scikit-learn: Machine Learning in Python", *J. Mach. Learn. Res.*, vol. 12, pp. 2825–2830, 2011.

[23] M. Stojkovic y J. Butt, "Industry 4.0 Implementation Framework for the Composite Manufacturing Industry", *J. Compos. Sci.*, vol. 6, n.o 9, p. 258, sep. 2022, doi: 10.3390/jcs6090258.

[24] "Tableau (version. 9.1)", *J. Med. Libr. Assoc. JMLA*, vol. 104, n.° 2, pp. 182, 183, abr. 2016, doi: 10.3163/1536-5050.104.2.022.

[25] Microsoft, *Power Bi Desktop* (2024). Microsoft.

[26] H. Castro et al., "Data science for Industry 4.0 and sustainability: A survey and analysis based on open data", *Machines*, 2023, vol. 11, p. 452, DOI: 10.3390/machines11040452

Index

www.ingramcontent.com/pod-product-compliance
Lightning Source LLC
Chambersburg PA
CBHW081256170526
45165CB00011B/3319